AF471199

Springer
Proceedings in Physics 39

Disorder and Nonlinearity

Proceedings of the Workshop
J. R. Oppenheimer Study Center
Los Alamos, New Mexico, 4–6 May, 1988

Editors: A. R. Bishop, D. K. Campbell,
and S. Pnevmatikos

With 66 Figures

Springer-Verlag Berlin Heidelberg New York
London Paris Tokyo Hong Kong

Dr. Alan R. Bishop
Dr. David K. Campbell
Dr. Stephanos Pnevmatikos
Los Alamos National Laboratory, Center of Nonlinear Studies
Los Alamos, NM 87545, USA

ISBN 3-540-51374-4 Springer-Verlag Berlin Heidelberg New York
ISBN 0-387-51374-4 Springer-Verlag New York Berlin Heidelberg

Preface

In the past three decades there has been enormous progress in identifying the essential role that "nonlinearity" plays in physical systems. Classical nonlinear wave equations can support localized, stable "soliton" solutions, and nonlinearities in quantum systems can lead to self-trapped excitations, such as polarons. Since these nonlinear excitations often dominate the transport and response properties of the systems in which they exist, accurate modeling of their effects is essential to interpreting a wide range of physical phenomena. Further, the dramatic developments in "deterministic chaos", including the recognition that even simple nonlinear dynamical systems can produce seemingly random temporal evolution, have similarly demonstrated that an understanding of chaotic dynamics is vital to an accurate interpretation of the behavior of many physical systems. As a consequence of these two developments, the study of nonlinear phenomena has emerged as a subject in its own right.

During these same three decades, similar progress has occurred in understanding the effects of "disorder". Stimulated by Anderson's pioneering work on "disordered" quantum solid state materials, this effort has also grown into a field that now includes a variety of classical and quantum systems and treats "disorder" arising from many sources, including impurities, random spatial structures, and stochastic applied fields.

Significantly, these two developments have occurred rather independently, with relatively little overlapping research. And yet in many if not most circumstances "disorder" and "nonlinearity" coexist, and their separate effects can reinforce, complement, or frustrate each other. A clear example is provided by "localization" phenomena in solids: just as "disorder" due to random potentials can lead to "Anderson localization" of electronic states, so "nonlinearities" arising from couplings between electrons and phonons can produce the localized electronic states corresponding to polarons. In a wide variety of materials, distinguishing effects arising from disorder and randomness from those due to nonlinearity has become an important issue in interpretation of experiments. In a like manner, the randomness of deterministic chaos can in many respects mimic that produced by stochastic external forces; again, unraveling the intertwined effects of stochasticity and chaos is an important issue in understanding many physical systems. Further, the recently established formal similarities between problems involving "quantum chaos" and Anderson localization indicate yet another aspect of the growing overlap of nonlinearity and disorder. Thus, quite generally modeling and interpreting the possible combined effects of nonlinearity and disorder remains an important area for future research.

The goal of the workshop on "Disorder and Nonlinearity in Classical and Quantum Systems", on which these proceedings are based, was to bring together researchers involved in these two presently disjoint research areas to discuss ways in which their separate expertise could be combined to confront common problems. Apart from opening channels of communication among researchers in the two areas, the workshop aimed to identify fundamental physical questions that involve both nonlinearity and disorder and to outline a research program to answer these questions. By all measures, the meeting, which was held in May 1988 at Los Alamos, was a success. It is our hope that the present proceedings will make this apparent to those among our readers (and, in some cases, authors(!)) who were not able to attend the event itself.

Finally, as members of the organizing committee, we wish to express our gratitude to those whose assistance helped insure the success of the workshop. Financial sponsorship came largely from the Applied Mathematical Sciences Program of the U.S. Department of Energy. Excellent administrative support was provided by Marian Martinez, Frankie Gomez, Dorothy Garcia, Lucille Martinez, and Valerie Ortiz, all of the Center for Nonlinear Studies. The Los Alamos National Laboratory contributed the outstanding conference facilities and organizational staff. And the contributions of the speakers and participants established that, to use the phrasing of Jim Krumhansl, "nonlinearity *with* disorder" will form a subject worthy of study for many years to come.

Los Alamos,
February 1989

A. Bishop
D. Campbell
S. Pnevmatikos

Contents

Part I

Localization and Nonlinearity

Nonlinearity and Localization in One-Dimensional Random Media

R. Knapp [1], *G. Papanicolaou* [2], *and B. White* [3]

[1] Institute for Mathematics and its Applications, University of Minnesota,
 514 Vincent Hall, 206 Church Street SE, Minneapolis, MI 55455, USA
[2] Courant Institute of Mathematical Sciences, New York University,
 251 Mercer Street, New York, NY 10012, USA
[3] Exxon Research and Engineering Company, Route 22 East,
 Annandale, NJ 08801, USA

A nonlinear Fabry-Perot etalon with random inhomogeneities is modeled by a one-dimensional stochastic Helmholtz equation. An asymptotic estimate of the threshold intensity needed for optical bistability is found for homogeneous media as a function of length. In the random case localization is affected and estimates of the energy growth are derived for large lengths with fixed output. Comparisons of the theory and numerical simulations are presented.

1 Introduction

In nonlinear optical media the intensity of light changes the index of refraction giving rise to many interesting phenomena, including optical bistability. When an optical device can have two different output states for a given input intensity (depending on hysteresis) it is said to be bistable. On the other hand in linear media with random inhomogeneities different phenomena arise, such as localization. When localization occurs, the transmitted intensity of an optical device decays exponentially as a function of the size of the material. Here we study nonlinear optical media with random inhomogeneities to gain some understanding of the interactions between disorder and nonlinearity and how these phenomena are affected.

Optical bistability was first observed experimentally using a Fabry-Perot interferometer or etalon [22]. More recently it has been achieved in semiconductors [14,23] and at room temperature [15]. The etalon is constructed from a slab of nonlinear material of thickness L sandwiched between two partially reflecting dielectric films. The medium in these experiments has a Kerr nonlinearity in which the potential depends on the intensity of light in the medium. Bistable optical devices can be designed to act as switches in which there is a threshold intensity. Above the threshold, transmission jumps to a higher state, or the device "switches on". Descriptions of applications and references are given in [31]. A nice physical explanation of optical bistability appears in [12]. Randomness can be introduced into the etalon either by introducing impurities into the nonlinear medium or by constructing the nonlinear medium with alternating films of linear and nonlinear media of random thickness. An important physical model of the nonlinearity in an etalon has been developed in [21].

We have simplified the problem by choosing to consider only time harmonic fields with a constant input intensity. The equations and formulation of the problem we study is described in section 2. In the nonlinear case the problem is complicated by the possibility of bistability. If there is bistability, the equations will not have a unique solution for a given input intensity. However, for a given output intensity, there are unique values of input intensity and the transmission coefficient. In section 5 we transform the boundary value problem with incoming, reflected and transmitted waves into an initial value problem depending on a parameter α equal to the output intensity. This is a useful way to study the boundary value problem but it is not as effective as in linear reflection-transmission problems.

Springer Proceedings in Physics, Vol. 39   **Disorder and Nonlinearity**
Editor: A.R. Bishop   © Springer-Verlag Berlin, Heidelberg 1989

A number of studies have been done when no random inhomogeneities are present. Bistability in the reflection coefficient for a semi-infinite medium has been studied fairly extensively including some cases with absorbtion which we do not consider [8,20]. When the etalon is of the order of the wavelength of the light at zero intensity some results are given in [8]. The authors use a transformation similar to the one in section 4, but we consider the case where the etalon length is large compared to the wavelength of the zero intensity light. In this section we show that in the asymptotic limit of large length,

$$w_c \sim \frac{C}{L^{\frac{1}{3}}}$$

where w_c is the threshold intensity needed for the onset of bistability.

In the random case less is known. Extensive studies have been done for linear media and the results are surveyed in section 3. A few studies have been done with nonlinear random media however [3,2,9,10,18]. In [10] the authors have shown, using the same equations we study, that for the fixed output problem in the nonlinear random Fabry-Perot etalon the transmittivity (the square modulus of the transmission coefficient) does decay with increasing length, but the rate is at most quadratic. This is in contrast to the exponential localization in the linear case. In [17] we extend and sharpen the results described in [10].

Sections 5-8 are devoted to the fixed output problem in or near the diffusion limit. We use a parameter ϵ which governs the size of the random perturbations and their length scales so that in the limit $\epsilon \to 0$ we have white noise. In section 6 we transform the initial value problem described in section 5 to action-angle variables (I, ϕ) and show how the new equations can be expressed in terms of Jacobian elliptic functions. In this form the equations can be scaled so that they can be analyzed asymptotically as we do in section 8. In the limit $\epsilon \to 0$ I is governed by a diffusion equation depending on the fixed output parameter α. In section 8 we also describe how first passage times from an interval $(0,M)$ in the limit $M \to \infty$ can be used to determine the growth rate of the process I. We show, using asymptotic expansions for large energy and action, that the growth satisfies

$$\lim_{t \to \infty} P_I \left\{ I(t) \le t^{\frac{3}{4}+\delta} \right\} = 1$$

$$\lim_{t \to \infty} P_I \left\{ I(t) \ge t^{\frac{3}{4}-\delta} \right\} = 1$$

for $\delta > 0$ where t is a scaled length. The $t^{3/4}$ growth of the action is equivalent to linear growth of the energy.

In [17] we address briefly the fixed input problem. Here we only mention our numerical results in section 9.

An important part of the research consists of numerical experiments. We have developed a code which uses elliptic functions to solve both quickly and accurately the equations for fixed output α. The code was quite successful in verifying the results for the deterministic problem (section 4) and for the random problem with large energy and length (section 8). Extensive numerical study has also been directed at understanding the fixed input problem in the random case. A brief description of the code and graphs of the results are given in section 9.

2 Formulation and Scaling

We will consider propagation of monochromatic waves in a one-dimensional, nonlinear, and randomly inhomogeneous medium. The time harmonic scalar field amplitude satisfies the equation

$$u_{xx} + k^2 n^2 \left(x, |u|^2 \right) u = 0, \quad x \in \mathbf{R}^1. \tag{1}$$

Here k is the free space wave number and $n \left(x, |u|^2 \right)$ is the index of refraction of the inhomogeneous and nonlinear medium. We assume that outside the interval $[0, L]$ the index of refraction is equal to one. We assume further that a plane wave of unit amplitude is incident from the right so that

$$u(x) = A_0 \left(e^{-ikx} + Re^{ikx} \right) \quad \text{for } x > L \tag{2}$$

and

$$u(x) = A_0 T e^{-ikx} \quad \text{for } x < 0. \tag{3}$$

The reflection and transmission coefficients are denoted by R and T, respectively, and A_0 is the amplitude of the incident wave. Equation (1) along with (2),(3), and the continuity of u and u_x at $x = 0$ and $x = L$ constitute the scattering problem that we shall analyze.

To model the nonlinear medium we will take the index of refraction n to have the form

$$n^2 \left(x, |u|^2 \right) = 1 + \mu(x) + \tilde{w} \, |u|^2 \tag{4}$$

with $\mu(x)$ a bounded, stationary, zero mean random process and $\tilde{w}$ a constant which can be positive or negative depending on the type of material that makes up the medium. Note that in the absence of nonlinearity, when $\tilde{w} = 0$, and of randomness, when $\mu(x) \equiv 0$, the index of refraction (4) is matched to that of the adjoining space $i.e.$ $n \equiv 1$. In most problems in nonlinear optics this is not an appropriate assumption. The base medium inside $[0, L]$ is mismatched with the adjoining media. Our assumption simplifies the analysis without changing qualitatively the results or requiring substantially different methods. With these assumptions, if the length is scaled by the wavenumber k and u is normalized by the input intensity, $|\mathbf{A}_0|$, the equations and boundary conditions can be written more simply as

$$\begin{aligned}
u_{xx} + k^2 \left(1 + \epsilon\mu(x) + w \, |u|^2 \right) u &= 0 \qquad 0 < x < L \\
u_x(0) + iu(0) &= 0 \\
u_x(L) - iu(L) &= -2e^{-iL}
\end{aligned} \tag{5}$$

where $w = \tilde{w} \, |\mathbf{A}_0|^2$. We will use these equations in the calculations that follow in later sections.

There are three length scales that arise in the scattering problem: the wavelength $\lambda = 2\pi/k$, the correlation length l of the random perturbation $\mu(x)$ and the size of the scattering region L. There are also two intensity parameters: the standard deviation σ of μ $\left(\sigma = (E\,\{\mu^2\})^{1/2} \right)$ and the size of the nonlinearity w. There are several asymptotic limits that can be analyzed depending on the relative sizes of these parameters. Let us choose units of length so that that $L = 1$. The simplest limit for (1)-(4) is the <u>effective medium</u> or <u>homogenization</u> limit in which $\lambda \sim 1$ and $l \sim \epsilon^2$ where ϵ is a small parameter. The intensity parameters σ and w are also of order one as ϵ tends to zero. In this very low frequency limit fluctuations are not important and the μ can be ignored in (4). This limit can be analyzed quite generally, not only in one space dimension [24].

Another limit that can be analyzed is when (with $L = 1$) $\lambda \sim \epsilon$ and $l \sim \epsilon^2$ while σ and w are of order one as ϵ tends to zero. In this limit fluctuations are important. In the linear case, $w = 0$, this scaling arises in geophysical applications, in underwater sound propagation, and elsewhere [6,27,28]. It has not been studied in the nonlinear case.

The limit that we will study here and which seems to be most appropriate for applications in optics is when (with $L = 1$) $\lambda \sim \epsilon^2$ and $l \sim \epsilon^2$ so that fluctuations are important. In fact, since the wavelength is comparable to the size of the inhomogeneities we should not expect to be able to analyze this limit unless the random perturbations are weak. So we take $\sigma \sim \epsilon$ and then $\epsilon\mu(x)$ behaves like white noise when $x \sim \epsilon^{-2}$. The size of the nonlinearity is of order one as ϵ tends to zero.

In all three limits loss of uniqueness (bistability) for the scattering problem (1)-(4) arises. We want to explore here how the value of w for which bistability first occurs depends on the other parameters of the problem, for example on L, the properties of μ, etc.

4

3 Linear Random Media and Localization

In this section we will briefly survey some well–known results about scattering by a slab of random medium in the linear case. The most important phenomenon in this case is localization of the scattering by the random inhomogeneities.

The complex–valued wave amplitude in monochromatic wave propagation satisfies the stochastic boundary value problem

$$
\begin{aligned}
u_{xx} + k^2 \left(1 + \mu(x)\right) u &= 0 \quad , \quad & 0 < x < L \\
u(x) &= e^{-ikx} + Re^{ikx} \quad , \quad & x > L \\
u(x) &= Te^{-ikx} \quad , \quad & x < 0.
\end{aligned}
\tag{6}
$$

with u and u_x continuous at $x = 0$ and $x = L$. The square of the refractive index has the form $1 + \mu(x)$ where $\mu(x)$ is a zero mean stationary and bounded, say $|\mu(x)| \leq 1/2$, real valued stochastic process. The reflection and transmission coefficients R and T are complex–valued random variables that depend on the wavenumber k, the width of the slab L and the random process $\mu(\cdot)$. Since $\mu(x)$ is real–valued, conservation of energy flux implies that

$$
|R|^2 + |T|^2 = 1
\tag{7}
$$

Localization means that the transmission coefficient decays exponentially with L, something that does not happen for periodic media or in three dimensions when the scattering inhomogeneities are weak. When $\mu(x)$ is a stationary ergodic random process with trivial tail σ-field then [7] there exists a positive finite number $l(k)$, the localization length, such that

$$
\lim_{L \to \infty} \frac{1}{L} \log |T|^2 = -\frac{2}{l(k)}
\tag{8}
$$

with probability one. The meaning of (8) is that the transmission coefficient decays exponentially

$$
|T| \sim e^{-\frac{L}{l(k)}}
\tag{9}
$$

for L large and in particular slabs significantly wider than the localization length allow very little transmission. The main ingredient in the proof of (8) is a version of Fürstenberg's theorem [13] for the linear random oscillator.

The dependence of the localization length on the wavenumber k or the noise intensity $\sigma = E\left\{\mu^2\right\}^{1/2}$ is complicated and not known in general. However it is easy to show (cf. [4] and references therein) that for k or σ small we have

$$
l(k) \sim \frac{\gamma}{k^2} \quad , \quad k \ll 1
\tag{10}
$$

and

$$
l(k) \sim \frac{\delta}{\sigma^2} \quad , \quad \sigma \ll 1
\tag{11}
$$

where γ and δ are positive constants. Combining these results with (8) (or the formal version (9)) we see that $|T|$ is likely to have a limit law as L tends to infinity provided $k \sim L^{-1/2}$ or $\sigma \sim L^{-1/2}$. In fact one can easily show [19] by the methods described in section 8 that R (or T) converge weakly in either of these limits to a diffusion process that have been studied extensively.

4 Homogeneous Nonlinear Media

In this section we study the case where the medium is homogeneous but nonlinear. This is the simplest case in which the effects of nonlinearity can be studied. We will derive an estimate of the intensity as a function of L for which bistability first occurs. Equations (5) with $w = 0$ will be used in the calculations.

The boundary value problem (5) can be converted to a family of initial value problems parametrized by the output intensity. Let

$$u(x) = \frac{q(x)}{|T|}e^{-i\theta(x)}. \tag{12}$$

We assume that the modulus of the transmission coefficient, $|T|$, is never zero. When (12) is put into (5) and the real and imaginary multiples of $e^{i\theta}$ are separated we get equations for θ and q. The boundary condition at $x = 0$ allows us to eliminate the θ equation and we are left with the q equation given by

$$q_{xx} - \frac{1}{q^3} + (1 + w|T|^2 q^2)q = 0 \tag{13}$$

with initial conditions

$$q(0) = 1$$
$$q_x(0) = 0.$$

This is an initial value problem for q. The quantity $w|T|^2$ can be thought of as a parameter in this formulation. Let $\alpha = w|T|^2$ so that $q = q(x; \alpha)$. We determine $|T|^2$ as a function of α by using the boundary condition at $x = L$. Taking the modulus of the boundary condition at $x = L$ in (5) and solving for $|T|^2$ gives

$$|T|^2(L, \alpha) = \frac{4}{q_x^2(L; \alpha) + q^2(L; \alpha) + 2 + \dfrac{1}{q^2(L; \alpha)}}$$

We now have $|T|^2$ as a function of α. Since $w = \alpha/|T|^2$ we also have w as a function of α. The quantity of principal interest in the scattering problem is $|T|^2$ as a function of w that is, the input–output power transfer relationship. In the present formulation we have obtained a parametric representation of this curve with α being the parameter. Since the $|T|^2$ $vs.$ w curve is not single valued when bistability occurs, the parametric representation of the curve is a necessary step in the study of bistability (see figure 1).

The expressions for $|T|^2$ and w can be simplified somewhat using the energy integral for (13)

$$q_x^2 + \frac{1}{q^2} + q^2 + \frac{1}{2}\alpha q^4 = 2 + \frac{\alpha}{2} \tag{14}$$

from which we deduce that q is periodic in x for each $\alpha \geq 0$. Recall we have restricted our analysis to consider only $\alpha \geq 0$. For small $\alpha < 0$, however, no fundamentally different analysis is necessary. Using the energy identity we can eliminate the derivative of q from the formulas for $|T|^2$ and w. The simpler formulas are given by

$$|T|^2(L, \alpha) = \frac{1}{1 + \frac{1}{8}\alpha(1 - q^4(L, \alpha))} \tag{15}$$
$$w(L, \alpha) = \alpha + \frac{1}{8}\alpha^2(1 - q^4(L, \alpha))$$

The solution of equation (13), being periodic, can be found explicitly in terms of Jacobian elliptic functions. In fact,

$$q(x;\alpha) = \sqrt{1 - (1 - e_3)\operatorname{sn}^2\left(-x\sqrt{\frac{\alpha}{2}(1 - e_3)};\, k^2\right)}. \tag{16}$$

Here $e_1 = 1, e_2$, and e_3 denote the roots of the cubic, $\frac{\alpha}{2}z^3 + z^2 - 2Ez + 1$, in descending order. The period of oscillation is $2K(k^2)/\sqrt{\frac{\alpha}{2}(1 - e_3)}$ where $K(k^2)$ is the complete elliptic integral of the first kind. The frequency ω is given by

$$\omega = \frac{2\pi}{\text{period}} = \frac{\sqrt{\frac{\alpha}{2}(1 - e_3)}}{K(k^2)} \tag{17}$$

It is clear from the formulation of the problem that both w and $|T|^2$ are single valued functions of the parameter α. Bistability occurs when $|T|^2$ ceases to be a single valued function of w. This occurs when $d|T|^2/dw$ becomes infinite. Since,

$$\frac{d|T|^2}{dw} = \frac{\dfrac{d|T|^2}{d\alpha}}{\dfrac{dw}{d\alpha}}$$

bistability will occur when

$$\frac{dw}{d\alpha} = 0. \tag{18}$$

The exact location of the first occurrence of bistability depends on whether the intensity is being increased or decreased. Figure 1 shows an example of a $|T|^2$ vs. w curve with the two points marked. If w is increased from zero, the curve will be followed until the point $(w_c^l, |T|^2(w_c^l))$ is reached. Beyond this point, if w is increased further, the value of $|T|^2$ will jump up to the curve above since it is a single valued function of w beyond w_c^l. A full analysis of this transition, however, requires study of the transient problem in addition to the time harmonic one. If from some point on the curve above the input intensity w is decreased, the curve will be followed until the point $(w_c^u, |T|^2(w_c^u))$ is reached. If w is decreased beyond this

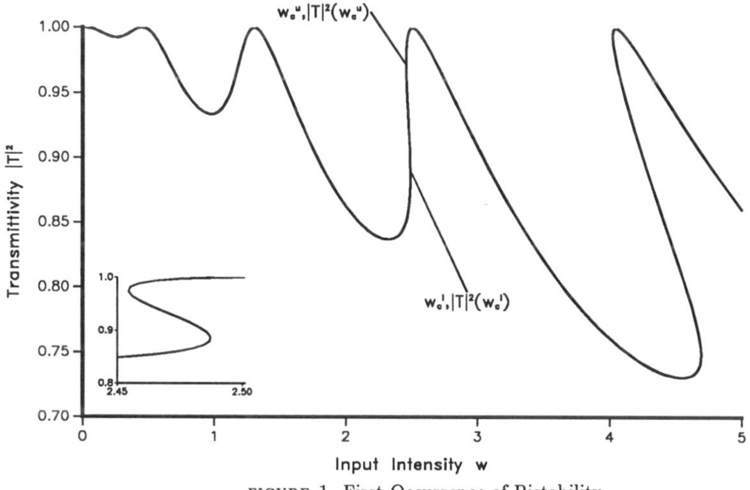

FIGURE 1. First Occurrence of Bistability

$|T|^2$ vs w curve for $L = 10$. The inset graph shows the region where bistability first occurs on an expanded scale for emphasis. α varies between 0 and about 4.3. $\alpha_c^l \simeq 2.02$ and $\alpha_c^u \simeq 2.3$.

point, the value of $|T|^2$ will jump down to the lower curve. Note that while w_c^l and w_c^u are very close, the corresponding values of transmittivity $|T|^2(w_c^l)$ and $|T|^2(w_c^u)$ and output intensity α_c^l and α_c^u are not close. When w is first increased beyond w_c^l and then decreased to below w_c^u the output as a function of input will trace a hysteresis loop typical of bistability. When w_c^l and w_c^u are close, the medium acts as an optical gate, where there is a threshold value for the higher transmission value. See [31,12] for description of these phenomena.

As the point w_c^l is approached from below, the rate of change of the output intensity becomes very large. This increase is due almost entirely to changes in the phase at $x = 0$ caused by the change in index of refraction due to the dispersive nonlinearity. The effects of length on the $|T|^2$ *vs.* w curve are shown in figure 2. When the medium is long, it requires less nonlinearity to bring about a change in phase since there are more wavelengths within the medium. Thus, for longer lengths, bistability will occur for smaller intensities and actually

$$\lim_{L \to \infty} w_c^u \le \lim_{L \to \infty} w_c^l = 0.$$

Clearly as w tends to zero, the corresponding values of α tend to zero. In what follows we find an expansion for $\alpha_c = \alpha_c^l$ and w_c^l as L tends to infinity.

We now expand everything for small α. The frequency ω defined in (17) has an expansion

$$\omega = \frac{\pi}{K(k^2)} \sqrt{\frac{\alpha}{2} (1 - e_3)} = 2 + \frac{3}{2}\alpha - \frac{21}{16}\alpha^2 + \frac{135}{64}\alpha^3 + \cdots.$$

As α approaches zero the parameter k^2 of the elliptic functions also approaches zero. This means that the Jacobian elliptic functions will approach sines and cosines. A change of variables from x to

$$y = -\omega x$$

makes the expansions of the elliptic function sn in terms of sines somewhat easier since (16) is π periodic in y and there are no secular terms in the expansion for small α given by

$$q^2 = 1 - \frac{1}{2}\alpha (1 - \cos y) + \frac{3}{4}\alpha^2 (1 - \cos y) - \frac{1}{32}\alpha^3 (45 - 44\cos y + 5\cos 2y) + \cdots$$

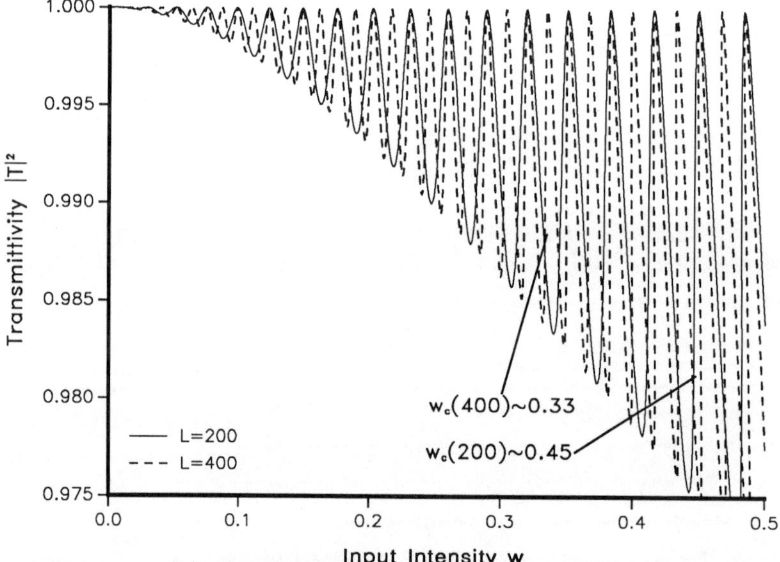

FIGURE 2. $|T|^2$ *vs.* w curve for $L = 200$ and $L = 400$

When these expansions are put into equation (15) for w we get

$$w = \alpha + \frac{1}{8}\alpha^3 \left(1 - \cos y\right) - \frac{1}{64}\alpha^4 \left(15 - 16\cos y + \cos 2y\right) + \frac{1}{128}\alpha^5 \left(63 - 68\cos y + 5\cos 2y\right) \cdots \tag{19}$$

where functions of y are evaluated at $y = -\omega L$. The α-derivative of w is given by

$$\frac{dw}{d\alpha} = 1 + \frac{1}{4}\alpha^3 \left[\sin y - \frac{\alpha}{4}\left(4\sin y - \sin 2y\right) + \frac{\alpha^2}{8}\left(34\sin y - 5\sin 2y\right)\right]\frac{dy}{d\alpha}$$
$$+ \frac{1}{8}\alpha^2 \left[3\left(1 - \cos y\right) - \frac{\alpha}{2}\left(15 - 16\cos y + \cos 2y\right) + \frac{5\alpha^2}{16}\left(63 - 68\cos y + 5\cos 2y\right)\right] + \cdots \tag{20}$$

with

$$\frac{dy}{d\alpha} = -L\frac{d\omega}{d\alpha} = -L\left(\frac{3}{2} - \frac{21}{8}\alpha + \frac{405}{64}\alpha^2 + \cdots\right)$$

To find α_c we set $dw/d\alpha = 0$ in (20) and divide both sides of the equation by L to get

$$-\frac{1}{L} = \frac{3}{16}\alpha_c^3\left[\sin y - \frac{\alpha_c}{4}(15\sin y - \sin 2y) + \frac{\alpha_c^2}{32}(383\sin y - 34\sin 2y)\right]$$
$$+ \frac{1}{8}\frac{\alpha_c^2}{L}\left[3\left(1 - \cos y\right) - \frac{\alpha_c}{2}\left(15 - 16\cos y + \cos 2y\right) + \frac{5\alpha_c^2}{16}\left(63 - 68\cos y + 5\cos 2y\right)\right] \tag{21}$$

The only term on the right hand side of equation (21) which can be comparable to $1/L$ is the term of order α_c^3. It is clear upon inspection that this term is larger than the term of order α_c^2/L as $L \to \infty$. Thus, to lowest order, α_c is given by the solution to

$$-\frac{1}{L\alpha_c^3}\frac{16}{3} = \sin\left(\left(2 + \frac{3}{2}\alpha_c + \cdots\right)L\right) \tag{22}$$

Consider each side of this equation to be a function of α_c with parameter L. Denote the two functions by $f_l(\alpha_c)$ and $f_r(\alpha_c)$. The first point of bistability, α_c^l is the smallest value of α_c at which the two curves are equal. The curve $f_l(\alpha_c)$ is a monotone increasing function of α_c. On the other hand, $f_r(\alpha_c)$ is oscillating between -1 and 1. The frequency of the oscillations increases with L. As $L \to \infty$ the oscillations become so fast that the curve $f_l(\alpha_c)$ will intersect nearly at the lower envelope of the oscillations of $f_r(\alpha_c)$. Thus $f_l(\alpha_c)$ is approximately equal to -1, so

$$\alpha_c = \left(\frac{16}{3}\right)^{\frac{1}{3}}\frac{1}{L^{\frac{1}{3}}} + \cdots \tag{23}$$

The result (23) can be stated more precisely as follows. There are constants C_1 and C_2, $C_1 > C_2 > 0$, independent of L such that

$$\limsup_{L\to\infty}\left[L\left(\alpha_c - \left(\frac{16}{3}\right)^{\frac{1}{3}}L^{-\frac{1}{3}} - \frac{5}{4}\left(\frac{16}{3}\right)^{\frac{2}{3}}L^{-\frac{2}{3}}\right)\right] = C_1$$

and

$$\liminf_{L\to\infty}\left[L\left(\alpha_c - \left(\frac{16}{3}\right)^{\frac{1}{3}}L^{-\frac{1}{3}} - \frac{5}{4}\left(\frac{16}{3}\right)^{\frac{2}{3}}L^{-\frac{2}{3}}\right)\right] = C_2.$$

The detailed proof is given in [17].

It is clear from the expression (19) for w that w_c^l has the same expansion as α_c up to third order. Thus,

$$w_c^l(L) = \left(\frac{16}{3}\right)^{\frac{1}{3}} L^{-\frac{1}{3}} + \frac{5}{4}\left(\frac{16}{3}\right)^{\frac{2}{3}} L^{-\frac{2}{3}} + \cdots. \tag{24}$$

The bistability point, w_c^u for decreasing input is $w(\alpha_c^u)$ where α_c^u is the next intersection of the curves in (22). Since $\alpha_c^u \geq \alpha_c^l$, but $w_c^u \leq w_c^l$, from (19) it is clear that the difference between α_c^l and α_c^u must be of order $1/L$, since if not, $w(\alpha_c^u) \geq w(\alpha_c^l)$ as $L \to \infty$. Thus, w_c^l and w_c^u are close to within third order also.

Since $1/L^{\frac{1}{3}}$ does not decrease very quickly with L the first term of the expansion (24) does not give a very good approximation to w_c except for extremely large L. Figure 3 shows that the second order term is sufficient to provide a very good approximation for moderate lengths L. The data was computed using a code in which the equation is solved exactly using the elliptic functions described above. The pattern that appears in the oscillations in the top part of Figure 3 is due to the way values of L were selected. In figure 4 we show the small scale

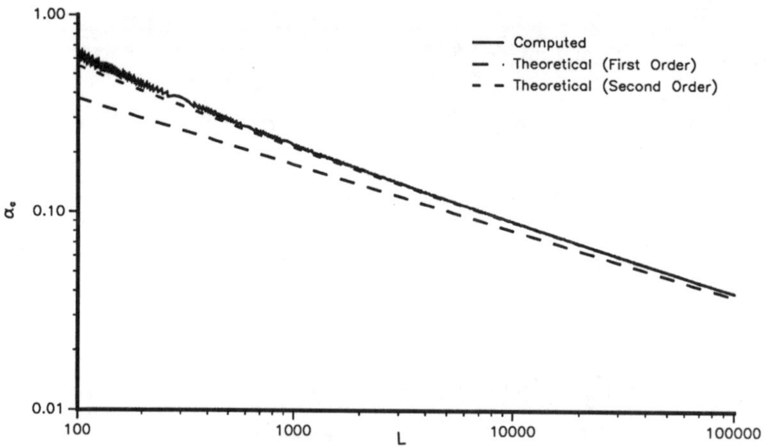

FIGURE 3. Computed and Theoretical values of α_c

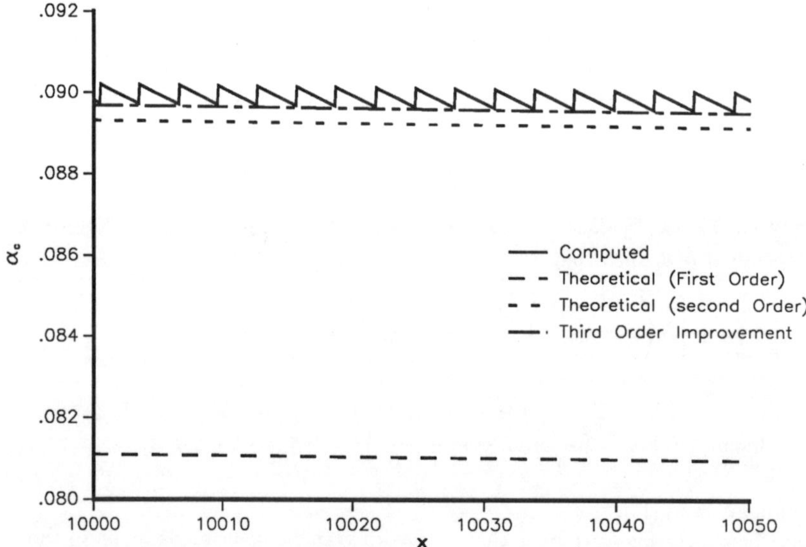

FIGURE 4. Oscillations of α_c as a Function of L

oscillations of α_c. The jumps in α_c occur when the number of complete periods within the medium for $\alpha = \alpha_c$ changes by one. On the $|T|^2$ vs w curve this corresponds to having the point $(w_c^l, |T|^2(w_c^l))$ move from the side of one transmission peak to the next. This corresponds to moving to the next bump on the $|T|^2$ vs. w curve. While the expansion will always have an error of third order $(1/L)$, this error can be minimized by making a third order correction to the envelope approximation. This is the third order correction shown in figure 4 which is right at the envelope of the oscillations.

5 Formulation of the Random Problem

In the remainder of this paper we consider the scaled nonlinear scattering problem (5) where $\mu(x)$ is a bounded $(|\mu(x)| \leq 1/2)$, zero mean $(E\{\mu(x)\} = 0)$, stationary random process. We introduce $\epsilon \ll 1$ a small parameter multiplying μ to control the size of the noise. We restrict the study for positive nonlinearity where $w \geq 0$.

We will use the transformation of the boundary value problem to an initial value problem as before and express the equations as a first order system.

$$
\begin{aligned}
q_x &= p & q(0;\alpha) &= 1 \\
p_x &= \frac{1}{q^3} - (1 + \epsilon\mu(x))q - \alpha q^3 & p(0;\alpha) &= 0
\end{aligned}
\tag{25}
$$

with $w|T|^2 = \alpha \geq 0$ a parameter. The transmittivity $|T|^2$ depends on the length L and the parameter α

$$
|T|^2(L,\alpha) = \frac{4}{p^2(L;\alpha) + \dfrac{1}{q^2(L;\alpha)} + 2 + q^2(L;\alpha)} = \frac{2}{E + 1 - \dfrac{1}{4}\alpha q^4}
$$

where E is the energy given by

$$
E = \frac{1}{2}\left(p^2(L;\alpha) + 1/q^2(L;\alpha) + q^2(L;\alpha) + \frac{1}{2}\alpha q^4(L;\alpha)\right).
\tag{26}
$$

The system (25) is Hamiltonian with x-dependent Hamiltonian

$$
H(q,p,x) = E(q,p) + \frac{1}{2}\epsilon\mu(x)q^2.
\tag{27}
$$

The parameter $\alpha = w|T|^2$ is the output intensity for the medium. For a fixed output α there are unique values of input intensity and transmittivity. Thus, solving the scattering problem for fixed α is equivalent to finding the input intensity and transmittivity required for a fixed output intensity. This is of interest in some situations, but it is the fixed input problem, in which the output intensity is considered a function of the input intensity, that is more natural. In the linear case these two problems are simply related, but when nonlinearity is present the fixed output problem cannot be simply transformed to the fixed input problem. One obvious difficulty is bistability, since there are several possible values of output for a given input. The solution(s) $|T|^2$ to the fixed point equation

$$
|T|^2 = \frac{4}{p^2(L;w|T|^2) + \dfrac{1}{q^2(L;w|T|^2)} + 2 + q^2(L;w|T|^2)}
\tag{28}
$$

with w given are possible solutions of the fixed input problem. If there is no bistability, (28) will have a unique solution. However, if bistability is present, there are at least two solutions.

We are concerned mainly with the study of the energy growth, but by expressing $|T|^2$ and w in terms of the energy, some information can be obtained about their behavior. The expressions for $|T|^2$ and w in terms of the energy and α are given by

$$|T|^2(L, \alpha) = \frac{2}{E(L; \alpha) + 1 - \frac{1}{4}\alpha q(L; \alpha)^4}$$

$$w = \frac{1}{2}\alpha \left(E(L; \alpha) + 1 - \frac{1}{4}\alpha q(L; \alpha)^4\right). \tag{29}$$

In the linear case where $\alpha = 0$, if the energy becomes large as L tends to infinity it is clear that $|T|^2$ must become small. However when α is positive and fixed the large L behavior of E is still interesting and we will focus most of our study on it.

We will actually study the large L behavior of E when ϵ is small and in more detail two particular cases: when $\alpha E \ll 1$ so that we are near the linear regime and when $\alpha E \gg 1$ which is the strongly nonlinear regime. These statements are made precise in the following sections.

6 Transformation to Action-Angle Variables

In this section we calculate a transformation of the equations (25) using action angle variables. They will be expressed in terms of Jacobian elliptic functions so that they can be studied in more detail in particular cases.

Define the potential V and energy E just as one would for the unperturbed system.

$$V(q) = \frac{1}{2}\left(\frac{1}{q^2} + q^2 + \alpha\frac{1}{2}q^4\right)$$

$$E = \frac{1}{2}\left(p^2 + \frac{1}{q^2} + q^2 + \alpha\frac{1}{2}q^4\right) = \frac{1}{2}p^2 + V(q) \tag{30}$$

This is not a conserved quantity because of the perturbation $\mu(x)$. Define the action I by the equation

$$I = \frac{1}{2\pi}\oint p\,dq = \frac{1}{2\pi}\oint \sqrt{2(E - V(q))}\,dq \tag{31}$$

in which E is fixed. For the unperturbed system the period of oscillation is given by

$$\text{period} = \frac{2\pi}{\omega} = \oint \frac{dq}{p} = \oint \frac{dq}{\sqrt{2(E - V(q))}} = 2\pi\frac{\partial I}{\partial E}. \tag{32}$$

Since I is an increasing function of E it can be inverted to give

$$E = E(I).$$

Once (31) is inverted, I should be considered an independent variable instead of E. In the equations that follow E represents $E(I)$. Note also that equation (32) gives a natural definition of the frequency $\omega(I)$

$$\omega = \frac{\partial E}{\partial I}. \tag{33}$$

The angle variable ϕ is defined from the action function

$$S(I, q) = -\int^q p\,dq. \tag{34}$$

by setting

$$\phi = \frac{\partial S}{\partial I}(I, q) = -\int^q \frac{\partial p}{\partial I}dq = -\frac{\partial E}{\partial I}\int^q \frac{\partial p}{\partial E}dq = -\omega(I)\int^q \frac{dq}{p}. \tag{35}$$

Define a canonical transformation by

$$q = Q(I, \phi)$$
$$p = P(I, \phi) = \left(\sqrt{2(E(I) - V(Q(I, \phi)))} \right) \tag{36}$$

where Q is obtained by solving (35) for q. We use the equation

$$E_x = pp_x + qq_x - \frac{q_x}{q^3} + \alpha q^3 q_x = -\epsilon \mu(x) q q_x = -\epsilon \mu(x) q p.$$

to find the differential equations satisfied by I and ϕ which are

$$I_x = \epsilon \mu(x) h_\phi(I, \phi)$$
$$\phi_x = -\omega(I) - \epsilon \mu(x) h_I(I, \phi) \tag{37}$$

where $h(I, \phi) = \frac{1}{2} Q^2(I, \phi)$ is the slowly varying part of the hamiltonian (27) transformed into the new variables

$$H(I, \phi) = E(I) + \epsilon \mu(x) h(I, \phi) = E(I) + \frac{1}{2} \epsilon \mu(x) Q^2(I, \phi). \tag{38}$$

The transformation and equations in the action angle variables can be expressed in terms of Jacobian elliptic functions. While the expressions appear cumbersome they simplify in the limits we are interested in studying later. The motion represented by (25) is oscillating so the roots of $2(E - V(q))$ determine the bounds of the motion. Using (30) and with E fixed we determine the roots of

$$q^2 p^2 = -\frac{1}{2} \alpha q^6 - q^4 + 2E q^2 - 1 = 0. \tag{39}$$

Denote the three roots for q^2 by e_1, e_2, and e_3 where $e_1 \geq e_2 \geq e_3$. These are functions of E and α only. The large E behavior of these roots is given in section 8. The simplest quantity to compute is the frequency given by

$$\omega = \frac{\pi}{K(k^2)} \sqrt{\frac{1}{2} \alpha (e_1 - e_3)}. \tag{40}$$

where K denotes the complete elliptic integral of the first kind and the parameter k^2 is given by

$$k^2 = \frac{e_1 - e_2}{e_1 - e_3}. \tag{41}$$

The integral defining ϕ in (35) is given by

$$\phi = -\omega(I) \int \frac{dq}{p} = \frac{\pi}{K(k^2)} \int \frac{dt}{\sqrt{(1 - t^2)(1 - k^2 t^2)}}$$

where $(e_1 - e_2) t^2 = e_1 - q^2$. The inversion of this integral is given by the Jacobian elliptic function sn in the same way the inverse would be given by the sine function if k^2 were zero. Thus we have

$$t = \text{sn}\left(\frac{K(k^2) \phi}{\pi}; k^2 \right)$$

and hence

$$Q(E, \phi) = \sqrt{e_1 - (e_1 - e_2) \, \text{sn}^2 \left(\frac{2K(k^2)}{\pi} \phi; k^2 \right)}$$

The action, I, is given by

$$I = \frac{(e_1 - e_2)\sqrt{\frac{1}{2}\alpha(e_1 - e_3)}}{\pi} \int_0^1 \frac{t^2\sqrt{(1-t^2)(1-k^2t^2)}}{\frac{e_1}{e_1 - e_2} - t^2}dt. \tag{42}$$

This integral can be expressed as the sum of elliptic integrals of the second and third kind. Unfortunantly it does not have a nice inverse like that defining ϕ. Because of this we will continue to use the notation $E = E(I)$ rather than explicitly stating the form of the I dependence at this point.

The function h appearing in the differential equations (37) is given by

$$h(I,\phi) = \frac{1}{2}Q^2(I,\phi) = \frac{1}{2}\left(e_1\mathrm{cn}^2\left(\frac{K(k^2)}{\pi}\phi;k^2\right) + e_2\mathrm{sn}^2\left(\frac{K(k^2)}{\pi}\phi;k^2\right)\right) \tag{43}$$

and so the Hamiltonian H is

$$H(I,\phi) = E(I) + \frac{1}{2}\epsilon\mu(x)\left(e_1\mathrm{cn}^2\left(\frac{K(k^2)}{\pi}\phi;k^2\right) + e_2\mathrm{sn}^2\left(\frac{K(k^2)}{\pi}\phi;k^2\right)\right). \tag{44}$$

The actual functions h_ϕ and h_I appearing in the equations (37) involve differentiation of sn and cn with respect to the parameter k and have simpler forms in the asymptotic limit studied below.

7 Diffusion Limit

We will now study the limit described in section 2 where the correlation length of the random perturbation $\mu(x)$ is of the same order as the free space wavelength and both are small compared to the length of the medium. We will use the parameter ϵ to scale the equations so that the correlation length is of order ϵ^2. Let t be defined by

$$x = \frac{t}{\epsilon^2} \tag{45}$$

and

$$\mu^\epsilon(t) = \mu\left(\frac{t}{\epsilon^2}\right).$$

The process μ^ϵ has a correlation length of order ϵ^2. Nearly all stochastic processes of interest in this problem can be expressed as projections to $\mathbf{R}^1$ of Markov processes on $\mathbf{R}^d$ for some $d > 1$. Thus, without loss of generality, we take μ itself to be Markov.

We want to use the asymptotic methods described in [24] when $\epsilon \to 0$ in the equations expressed in action-angle variables (37). Since the ϕ equation has the quickly varying part $\omega(I)$ we introduce two new variables τ and ψ by

$$\phi = \tau + \psi$$

with

$$\tau_x = -\omega(I)$$

When the system with these new variables is expressed in terms of the scaled t the equations (37) become

$$\dot{\tau}^\epsilon(t) = -\frac{1}{\epsilon^2}\omega(I^\epsilon(t))$$

$$\dot{I}^\epsilon(t) = \frac{1}{\epsilon}\mu^\epsilon(t)h_\phi(I^\epsilon(t),\tau^\epsilon(t) + \psi^\epsilon(t)) \tag{46}$$

$$\dot{\psi}^\epsilon(t) = -\frac{1}{\epsilon}\mu^\epsilon(t)h_I(I^\epsilon(t),\tau^\epsilon(t) + \psi^\epsilon(t))$$

Since μ is Markov, the joint process $(\mu^\epsilon, \tau^\epsilon, I^\epsilon, \psi^\epsilon)$ is also Markov. Note that the "fast" angle variable τ has an equation with an $O\left(1/\epsilon^2\right)$ term on the right. An order ϵ^2 change in t thus brings an order one change to τ^ϵ and the argument of μ. Thus both the period and the correlation length are of order ϵ^2.

We pass to the backward Kolmogorov equation associated with (46). For f a bounded smooth function $f : [-1/2, 1/2] \times \mathbf{R}^+ \times \mathbf{R}^+ \times [0, 2\pi] \to \mathbf{R}$ let

$$u^\epsilon(t, \mu, \tau, I, \psi) = E\left\{ f\left(\mu^\epsilon(t), \tau^\epsilon(t), I^\epsilon(t), \psi^\epsilon(t)\right) | \mu^\epsilon(0) = \mu, \tau^\epsilon(0) = \tau, I^\epsilon(0) = I, \psi^\epsilon(0) = \psi \right\}.$$

which is the conditional expectation of f evaluated at the state at time t given the state at time $t = 0$. Then u^ϵ satisfies the equation

$$\frac{\partial u^\epsilon}{\partial t} = \mathcal{L}^\epsilon u \quad ; \qquad u^\epsilon(0, \mu, \tau, I, \psi) = f(\mu, \tau, I, \psi) \tag{47}$$

where

$$\mathcal{L}^\epsilon = \frac{1}{\epsilon^2}\left(Q - \omega(I)\frac{\partial}{\partial \tau}\right) + \frac{1}{\epsilon}\mu\left(h_\phi(I, \tau + \psi)\frac{\partial}{\partial I} - h_I(I, \tau + \psi)\frac{\partial}{\partial \psi}\right). \tag{48}$$

In the limit $\epsilon \to 0$, as in [24], we get a diffusion equation

$$\frac{\partial u}{\partial t} = \mathcal{L}u$$

where

$$\mathcal{L} = \frac{\partial}{\partial I}a(I)\frac{\partial}{\partial I} - \frac{\partial}{\partial I}b(I)\frac{\partial}{\partial \phi} - \frac{\partial}{\partial \psi}b(I)\frac{\partial}{\partial I} + \frac{\partial}{\partial \psi}c(I)\frac{\partial}{\partial \psi} \tag{49}$$

and

$$a(I) = \int_0^\infty R(s)\langle h_\phi(I, \tau + \psi)h_\phi(I, \tau + \psi + \omega(I)s)\rangle \, ds$$

$$b(I) = \int_0^\infty R(s)\langle h_\phi(I, \tau + \psi)h_I(I, \tau + \psi + \omega(I)s)\rangle \, ds \tag{50}$$

$$c(I) = \int_0^\infty R(s)\langle h_I(I, \tau + \psi)h_I(I, \tau + \psi + \omega(I)s)\rangle \, ds$$

where $\langle\cdot\rangle$ represents averaging over a period in τ. The Fokker-Plank equation for the limiting diffusion process $(I(t), \psi(t))$ is the same as (49) since the operator $\mathcal{L}$ is self adjoint.

8 Growth Rate of the Energy when the Energy is Large

In this section we will examine the form the diffusion coefficients (50) have when the action or the energy is large.

For large E, and α fixed, the roots of the cubic (39) have expansions of the form

$$e_1 = \frac{2}{\sqrt{\alpha}}\sqrt{E} - \frac{1}{\alpha} + O\left(\frac{1}{\sqrt{E}}\right)$$

$$e_2 = \frac{1}{2E} + O\left(\left(\frac{1}{\sqrt{E}}\right)^3\right)$$

$$e_3 = -\frac{2}{\sqrt{\alpha}}\sqrt{E} - \frac{1}{\alpha} + O\left(\frac{1}{\sqrt{E}}\right)$$

In this limit the frequency, ω, from (40) is given by

$$\omega = \frac{\sqrt{2}\pi}{K(\frac{1}{2})}\alpha^{\frac{1}{4}}E^{\frac{1}{4}}\left(1 + O\left(\frac{1}{\sqrt{E}}\right)\right)$$

When the energy and action are large, constant terms will make no appreciable difference, so we will find the action by integrating the reciprocal of the frequency ω.

$$I = \int \frac{\partial I}{\partial E} dE = \int \frac{1}{\omega} dE = \frac{K(\frac{1}{2})}{\sqrt{2}\pi} \frac{1}{\alpha^{\frac{1}{4}}} \int \frac{dE}{E^{\frac{1}{4}}} = \frac{2\sqrt{2}K(\frac{1}{2})}{3\pi} \frac{E^{\frac{3}{4}}}{\alpha^{\frac{1}{4}}} \left(1 + O\left(\frac{1}{\sqrt{E}}\right)\right)$$

We need to invert this expression to get $E = E(I)$ as an expansion in some appropriate power of I. Since the first term is $E = cI^{4/3}$, expansions in $\sqrt{E}$ will be equivalent to expansions in powers of $I^{2/3}$. Thus we can write the energy in terms of the action as follows

$$E(I) = C_E^2 \alpha^{\frac{1}{3}} I^{\frac{4}{3}} \left(1 + O\left(I^{-\frac{2}{3}}\right)\right)$$

and

$$\omega(I) = \frac{4}{3} C_E^2 \alpha^{\frac{1}{3}} I^{\frac{1}{3}} \left(1 + O\left(I^{-\frac{2}{3}}\right)\right)$$

where

$$C_E = \frac{1}{2}\left(\frac{3\pi}{K(\frac{1}{2})}\right)^{\frac{2}{3}}.$$

The function $h(I,\phi)$ from (43) which appears in the differential equations for I and ϕ is given by

$$h(I,\phi) = \frac{1}{2}\left(2C_E \frac{I^{\frac{2}{3}}}{\alpha^{\frac{1}{3}}} \operatorname{cn}^2\left(\frac{2K(k^2)}{\pi}; k^2\right) + \frac{1}{2C_E^2} \frac{I^{-\frac{4}{3}}}{\alpha^{\frac{1}{3}}} \operatorname{sn}^2\left(\frac{2K(k^2)}{\pi}; k^2\right)\right)$$

$$= C_E \frac{I^{\frac{2}{3}}}{\alpha^{\frac{1}{3}}} \operatorname{cn}^2\left(\frac{2K(k^2)}{\pi}; k^2\right) + O(1)$$

The ϕ-derivative of h is thus

$$h_\phi = 2C_E \frac{I^{\frac{2}{3}}}{\alpha^{\frac{1}{3}}} \frac{K(\frac{1}{2})}{\pi} \operatorname{sn}\left(\frac{1K(\frac{1}{2})}{\pi}\phi; \frac{1}{2}\right) \operatorname{cn}\left(\frac{1K(\frac{1}{2})}{\pi}\phi; \frac{1}{2}\right) \operatorname{dn}\left(\frac{1K(\frac{1}{2})}{\pi}\phi; \frac{1}{2}\right) + O(1)$$

and the I-derivative is given by

$$h_I = \frac{1}{3} C_E \frac{I^{-\frac{1}{3}}}{\alpha^{\frac{1}{3}}} \operatorname{cn}^2\left(\frac{1K(\frac{1}{2})}{\pi}\phi; \frac{1}{2}\right) + O\left(\frac{1}{I}\right).$$

It can be shown [17] using this that both h_ϕ and h_I have I-derivatives which are periodic and of order $I^{-\frac{5}{3}}$, which are negligible compared to the I-derivatives of the other terms in h_I and h_ϕ.

To compute the diffusion coefficients (50) we assume for simplicity that the correlation function $R(s)$ has the form

$$R(s) = C_R e^{-\gamma s}.$$

The coefficient $a(I)$ is given by

$$a(I) = \frac{A}{\alpha^{\frac{4}{3}}} I^{\frac{2}{3}} + O(1)$$

where the constant A can be computed explicitly [17]. The coefficient $a(I)$ would have been larger if ω did not grow with I. This will have the net effect of speeding up the growth rate of the process I. It can also be shown the $b(I)$ is order 1 and $c(I)$ is order $I^{-\frac{1}{3}}$ which are negligible compared to $a(I)$ for large I. Thus the limit diffusion process is governed by the generator

$$\mathcal{L} = \frac{\partial}{\partial I} \frac{A}{\alpha^{\frac{4}{3}}} I^{\frac{2}{3}} \frac{\partial}{\partial I}. \tag{51}$$

We will now use first passage times to obtain a formal estimate of the growth rate of the action. Let u_M be the mean exit time of $I(t)$ from the interval $(0, M)$ on the I-axis. Then it is well known that

$$\mathcal{L}u_M(I) = -1 \quad ; \qquad u_M(M) = 0. \tag{52}$$

Suppose that there is some function $f(M)$, $M \geq 0$, such that

$$\lim_{M \to \infty} \left(\frac{u_M(I)}{f(M)} \right) = 1 \tag{53}$$

and that the fluctuations about the mean go to zero as $M \to \infty$. We know by the definition of first exit time that

$$I(\tau_M) = M. \tag{54}$$

Think now of τ_M as a regular time, t, so that (54) is $I(t) = M$. Think also of M as a function of t–that is the value of I reached after a time t. But from (53) we have (assuming the fluctuations go to zero)

$$t \sim f(M).$$

This can be inverted since $f(M)$ is an increasing function of M, so

$$M \sim f^{-1}(t).$$

Putting this into (54) gives us the estimate for the growth rate of the process I

$$I(t) \sim f^{-1}(t), \quad t \to \infty. \tag{55}$$

The solution of (52) for the generator (51) is given by

$$u_M(I) = \frac{3\alpha^{\frac{4}{3}}}{4A} \left(M^{\frac{4}{3}} - I^{\frac{4}{3}} \right). \tag{56}$$

As $M \to \infty$ the dependence on I will be negligible so we can invert the exit time to get the growth rate of I using (55). The result is that

$$I(t) \sim \frac{1}{\alpha} \left(\frac{4}{3} At \right)^{3/4}, \quad \text{as } t \to \infty.$$

Note that the $t^{\frac{3}{4}}$ growth rate of the action I corresponds to linear growth of the energy E. In the linear case described in section 3 the fluctuations of the exit time τ_M from the mean u_M do die out and this process can be used to demonstrate the exponential behavior described in section 3. However in the nonlinear case the fluctuations do not die out [17]. The inversion gives a formal indication of the growth, but is not rigorous and (55) holds only in a weak form. The asymptotic estimate

$$I(t) \sim ct^{3/4}$$

is valid in the sense that

$$\lim_{t \to \infty} P_I \left\{ I(t) \geq ct^{3/4-\delta} \right\} = 1$$
$$\lim_{t \to \infty} P_I \left\{ I(t) \leq ct^{3/4+\delta} \right\} = 1 \tag{57}$$

for any $\delta > 0$. In the same sense as (57) the energy $E(t)$ grows like t as $t \to \infty$ with $\alpha > 0$ fixed.

9 Numerical Experiments

The diffusion limit has been extensively modeled using numerical experiments. We solved numerically the initial value problem

$$q_x = p \qquad\qquad q(0) = 1$$
$$p_x = \frac{1}{q^3} - (1 + \epsilon\mu(x))q - \alpha q^3 = 0 \qquad p(0) = 0$$

where the stochastic process μ was modeled as follows. On intervals of random length l_i, μ takes on a random value, μ_i distributed uniformly on $[-1, 1]$. The random lengths l_i are given by

$$l_i = -l\log(1 - z)$$

where z is distributed uniformly on $[0, 1]$ so they have an exponential law with mean l.

For the growth of the energy to stabilize in the nonlinear case very large values of L are necessary. Thus the code used to solve the equations needs to be both accurate and efficient. We found that when standard ODE solvers were used (LSODE and fourth order Runge-Kutta) the computed solutions did not converge even at moderate values of L when the error tolerance or step size was decreased. The solution over each random interval needs to be extremely accurate since the growing energy tends to magnify errors. Since these methods were inadequate we developed a code which solves the equations directly using Jacobian elliptic functions.

Over each random interval l_i, since μ is constant, the equation is an unperturbed oscillator. Suppose $q(x)$ and $p(x)$ are the solution at x which is at the end of the random interval l_{i-1} We want to solve the problem

$$q_y = p \qquad\qquad q(0) = q(x)$$
$$p_y = \frac{1}{q^3} - \nu q - \alpha q^3 \qquad p(0) = p(x) \tag{58}$$

at $y = l_i$ where $\nu = 1 + \epsilon\mu_i$. The solution can be expressed in terms of elliptic integrals and functions which can be numerically evaluated using the algorithm given in [26]. In the linear case and for extremely small values of α the transformation to elliptic functions becomes singular so a first order approximation (in α) was used for $\alpha < 10^{-4}$ [17]. The value at the end of the next random interval (or anywhere up to it) can thus be found given only $\alpha, \mu, q(x), p(x)$, and l_i. The solution of the problem for large L was achieved through repeated iterations of this routine for the randomly chosen intervals l_i and random values μ_i.

Figure 5 shows the mean energy as a function of x (or L) for several different values of α with ϵ held fixed. On first inspection the curves appear to be basically linear from the origin but, in fact, the growth does not stabilize to linear until x is quite large. This can be observed by dividing the energy by x as in figure 6. Note that the smaller α is the longer it takes for the growth to stabilize. Figure 7 shows the mean energy as a function of x for several values of ϵ with a fixed value of α. The linear growth rates acheived towards the right hand sides of figures 5 and 7 agree closely with the theoretically predicted result

$$E \simeq .03113\frac{\epsilon^2}{\alpha l}x$$

over a wide range of the parameters α and ϵ [17].

The most interesting dependence on α is in its effect on the transition region from the exponential growth of the linear case to the linear growth of the nonlinear case. Figure 8 shows the growth for α varying over four orders of magnitude and for the linear case. Some qualitative statements can be made based on the data in these figures. For $\alpha < 1$, the growth starts out as it does for the linear problem, but as the energy starts to become significant compared to α the growth slows. The initial slowdown, where the nonlinear curves first pulls away from the linear curve, seems to occur roughly when $E^2\alpha$ is a litle bit smaller than 1 (in fact, the range seems to vary between $.1 < E^2\alpha < .4$ for these runs.) By the time the energy is large and the growth has almost completely stabilized to linear, $E\alpha$ is large. The transition region lasts over a quite a long interval in x. The curves are spaced evenly with the logarithmic axis because the values of $\log\alpha$ are evenly spaced.

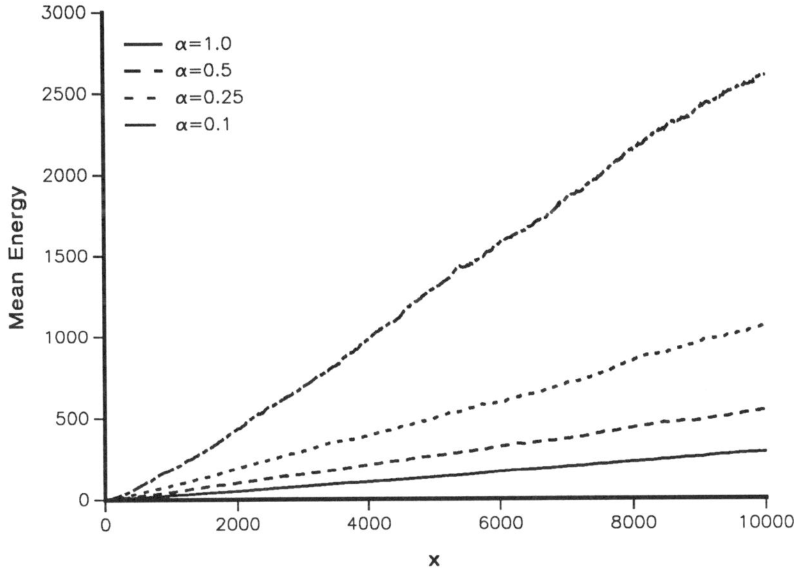

FIGURE 5. Mean Energy Growth for Different Values of α with Fixed ϵ

The curves represent the average energy for 1000 realizations with $\epsilon = 0.7$ and mean correlation length $l = 0.5$.

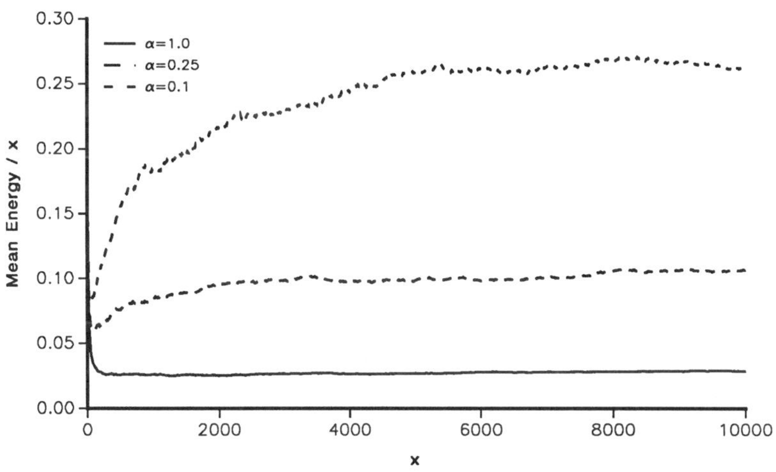

FIGURE 6. Stabilization to Linear Growth for Different Values of α

The curves represent the average energy divided by x for 1000 realizations with $\epsilon = 0.7$ and mean correlation length $l = 0.5$. The height of the flat part of each curve does not represent the asymptotic energy growth rate for that value of α because for these values of x, the intercept is still significant.

The computations described above all address the fixed output (α fixed) problem. Solving the fixed input problem for a given $w = w_0$ requires finding one of the solutions of equation (28). The simplest way to do this is using an iterative method. One value of the transmittivity, $|T|^2(L, w_0)$ is given by $\alpha/w(L, \alpha^*)$ where α^* is a root of the function $f(L, \alpha) = w(L, \alpha) - w_0$ with $w(L, \alpha)$ given by

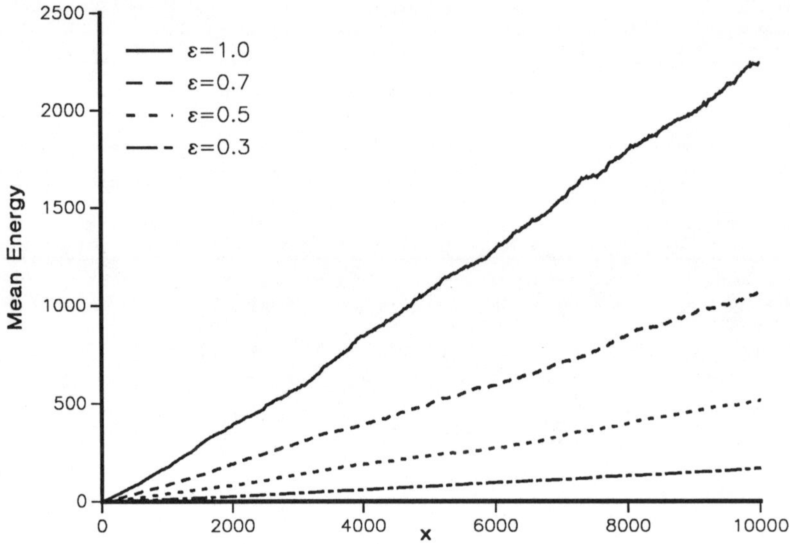

FIGURE 7. Mean Energy Growth for Different Values of ϵ with Fixed α

The curves represent the average energy for 1000 realizations with $\alpha = 0.25$ and mean correlation length $l = 0.5$.

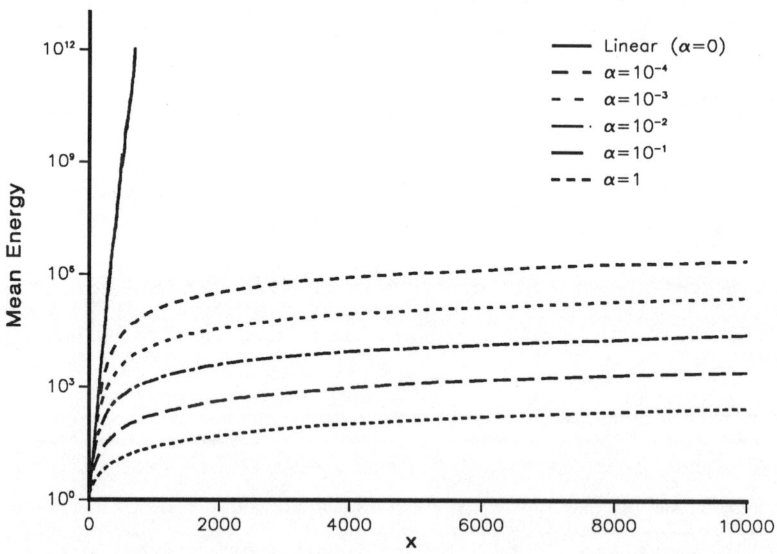

FIGURE 8. Comparison of Energy Growth with the Linear Case for Small x

The curves represent the average energy for 1000 realizations with $\epsilon = 0.7$ and mean correlation length $l = 0.5$. The vertical axis is scaled logarithmically so that the curves can be compared.

20

$$w(L,\alpha) = \frac{1}{4}\alpha \left(p(L,\alpha)^2 + q(L,\alpha)^2 + 2 + 1/q(L,\alpha)^2 \right).$$

Given an initial guess, α_0, the root can be found using an iterative method such as the secant method evaluating the function at each α using the code described above. An iterative method has at least two major drawbacks. First, there may be bistability and hence several solutions to the fixed point problem and it may not be possible to get the iterative method to converge to a particular one, such as $\max(|T|^2(w_0))$. In many realizations, the values of $|T|^2(w_0)$ are very close together and are difficult to distinguish. From realization to realization there is a large variation in the point at which bistability occurs and the iterative method finds only one solution at a time, so it is difficult to check for bistability. Second, since the values of α^* can vary a lot from realization to realization, it is hard to find a good initial value to start the iteration from. Thus for each value of w several iterations may be required.

One way to avoid these problems is to find the entire $|T|^2$ vs. w curve by computing $w(L,\alpha)$ and $|T|^2(L,\alpha)$ for $\alpha = \alpha_k$, with $0 = \alpha_0 < \alpha_1 < ... < \alpha_p = w_0$. The intersections of this curve with the line $w = w_0$ give all the possible values of $|T|^2(w_0)$ since $w = \alpha/|T|^2 \geq \alpha$. This method can be extended to find the values of $|T|^2(w_i)$ at a discrete set of values w_i [17]. Since it is not practical to store all the values for each w_i we have kept the number of crossings at $w = w_i$,

$$\max_{w=w_i} |T|^2(w) \quad \text{and} \quad \max_{w<w_i} |T|^2(w),$$

and the corresponding minima. The compiled results are averaged over many realizations.

Figure 9 shows the average minimum and maximum values of $|T|^2(w_i)$. No observable trends appear in the data until the appearance of bistability which is where the solid and dashed curves diverge. The increase in the maximum is natural to expect since when there is bistability there must be a higher and lower state. The consistent decrease in the minimum is a bit more surprising since this decreases form the exponentially localized linear value. Before bistability occurs the data is too oscillatory to have any real meaning. Figures 10 and 11 show the average lower and upper envelopes for the values of $|T|^2$ as a function of w for different fixed values of L. Figure 12 shows the same data plotted as a function of L for different values of w. The lower envelope tends to decrease consistently with both w and L, the rate of decrease

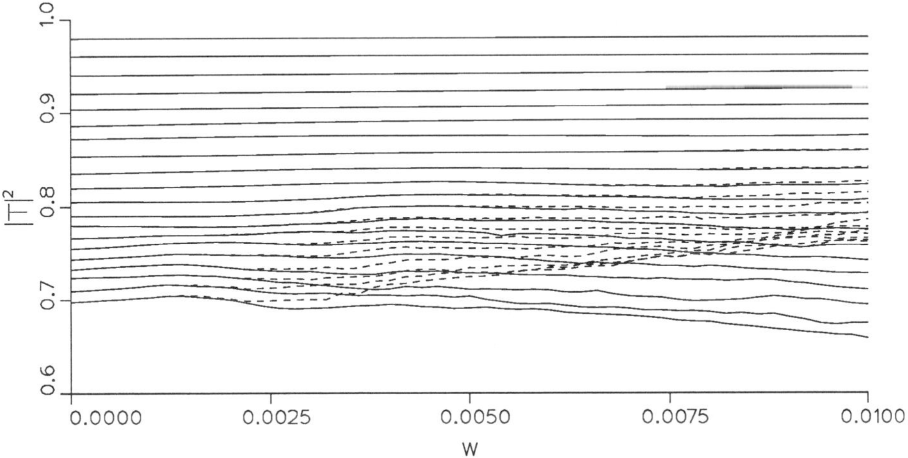

FIGURE 9. Average $|T|^2$ vs. w curves for small w

The curves represent the average value of $\max(|T|^2)$ and $\min(|T|^2)$ over 1000 realizations with $\epsilon = 0.1$ and $l = 0.5$. The ten solid curves from top (coincides with the axis) to bottom are for $L = 100, 200, \cdots, 1000$ respectively.

increasing with larger L. On the other hand, the upper envelope tends towards one value near $|T|^2 = 1$ as w increases independent of L. Thus, with nonlinearity present, the transmission coefficient is likely to have a maximum value near one. However at the same time it is likely to have a minimum smaller or near the exponentially localized linear value. Note that these represent average values and an individual realization may vary beyond these envelopes. It is interesting to compare the average variations in the random problem with those displayed in 2 for the homogeneous problem. For w ten times greater than the largest values displayed in

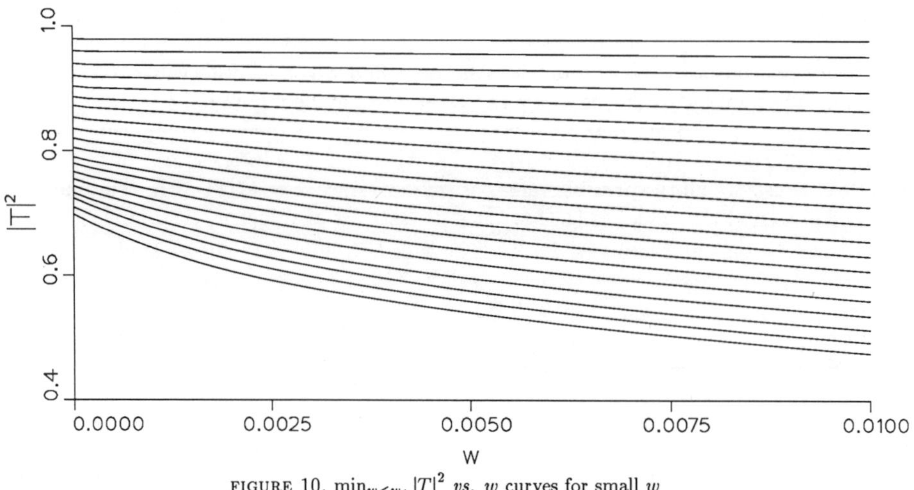

FIGURE 10. $\min_{w<w_i} |T|^2$ *vs.* w curves for small w

Each curve represents the average lower bound of $|T|^2$ for the plotted values of w_i for a fixed L. The twenty curves are for L=50 to 1000 in steps of 50 from top to bottom. Averages are taken over 1000 realizations with $\epsilon = 0.1$ and $l = 0.5$.

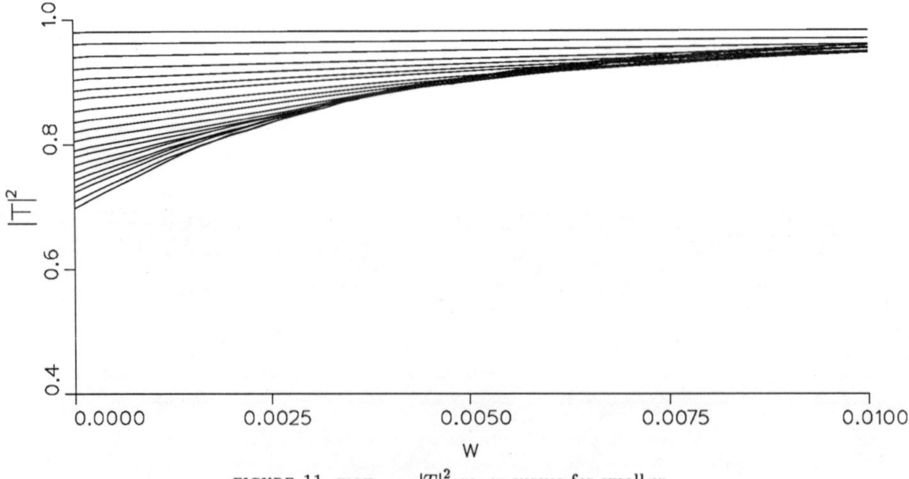

FIGURE 11. $\max_{w<w_i} |T|^2$ *vs.* w curves for small w

Each curve represents the $\min_{w<w_i}$ for the plotted values of w_i for a fixed L. The twenty curves are for L=50 to 1000 in steps of 50 from top to bottom. Averages are taken over 1000 realizations with $\epsilon = 0.1$ and $l = 0.5$.

figures 10-12 the variation between the minimum and maximum values of $|T|^2$ is roughly 0.0025 which is less than one hundreth of that for the random problem.

Figure 13 shows the level curves of the average number of values of $|T|^2$ as a function of L and w. Since the number of values will only be larger than one if there is bistability, the first curve gives a rough estimate of w_c, the first w for which bistability is likely to occur. Note that the first labeled curve shown (1.5 values on the average) represents the values of L and w for which there has been bistability for roughly one-fourth of the realizations. When this is compared to figure 3 it is readily apparant that the threshold intensity needed for bistability,

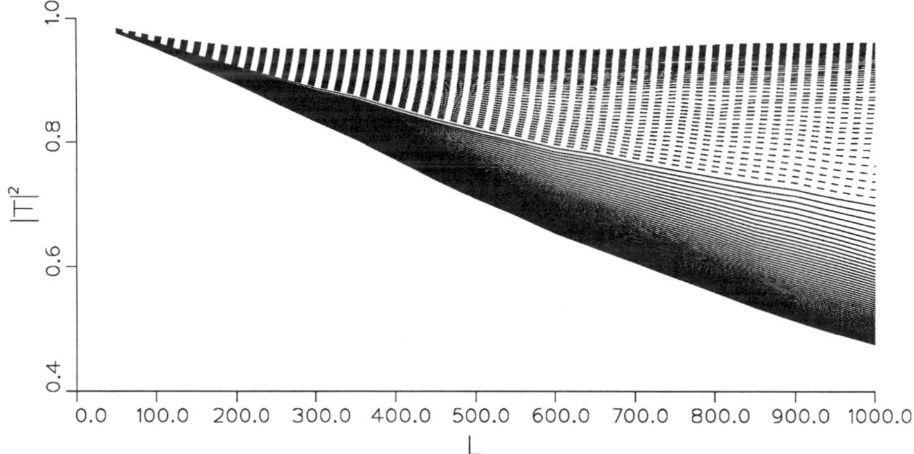

FIGURE 12. $min_{w<w_i}$ and $max_{w<w_i} |T|^2$ *vs.* L curves.

Each curve represents the $min_{w<w_i}$ for a fixed w_i plotted as a function of L. The solid curves are the average lower bound for w_i=0.0 to 0.01 in steps of 0.002 from top to bottom. The dashed curves are the average upper bound for w_i=0.0 to 0.01 in steps of 0.002.. Averages are taken over 1000 realizations with $\epsilon = 0.1$ and $l = 0.5$.

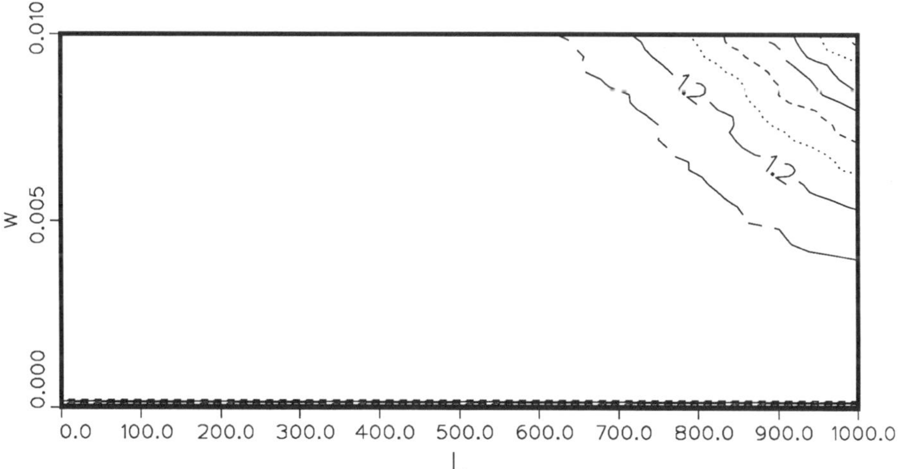

FIGURE 13. Level Curves of Number of $|T|^2$ Values

The level curves represent the average number of values of $|T|^2$ as a function of L and w for 1000 realizations with $\epsilon = 0.1$ and $l = 0.5$.

w_c, is much smaller when randomness is present. For example, $\epsilon = .1$ and $L = 1000$, bistability occurs for about 25 percent of the realizations when $w \simeq 0.008$, but in the homogeneous problem bistability does not occur for $L = 1000$ until $w \simeq 0.3$.

Some analytic study of the fixed point problem is given in [17]. These numerical results give an indication of some of the interesting phenomena which occur with the fixed input problem and should help to direct future research.

Acknowledgments

The work of R. Knapp and G. Papanicolaou was supported by the National Science Foundation under grant # NSF-DMS-870195 and by the Air Force Office of Scientific Research under grant # AFOSR-86-0352 .

References

[1] Abramowitz, M. and Stegun, I. (1965),
<u>Handbook of Mathematical Functions</u> Dover Publications, Inc., New York

[2] Al'tschuler, G.B., Ermolaev, V.S. (1982),
"Enhanced Transmission Due to Nonlinear Light Scattering by Static Optical Nonuniformities", *Sov. Phys. Dokl.* **28** 2 p146, American Institute of Physics

[3] Al'tschuler, G.B., Ermolaev, V.S., Krylov, K.I., Manenkov, A.A. (1983),
"Effects of Light Scattering in Nonuniform Media with Kerr Nonlinearity", *Sov. Phys. Dokl.* **28** 11 p951, American Institute of Physics

[4] Arnold, L. , Papanicolaou, G. , and Wihstutz, V. (1986),
"Asymptotic Analysis of the Lyapunov Exponent and Rotation Number of the Random Oscillator and Applications." *Siam Journal of Applied Math*, **28** 3

[5] Blankenship, G. and Papanicolaou G. C (1978),
"Stability and Control of Stochastic Systems with Wide-Band Noise Disturbances," *Siam Journal of Applied Math*, **34** 3 p347-476

[6] Burridge R., Papanicolaou, G., and White, B. (1987),
"Statistics for Pulse Reflection From a Randomly Layered Medium", *Siam J. Appl. Math.*,**47** p146-168

[7] Carmona, R. (1985),
"The Random Schrödinger Equation", Ecole d'Ete de Saint-Four, ed. P. Hennegrenin, Springer Lecture Notes in Mathematics, 1985

[8] Chen, W. and D.L. Mills (1987),
"Optical Response of a Nonlinear Dielectric Film", *Physical Review B* **35** 2 p524,

[9] Danileiko Y.K., Manenov A.A., Nechitailo V.S., Khaimov-Mal'kov V.Y. (1971),
"Nonlinear Scattering of Light in Inhomogeneous Media", *Sov. Phys. JETP* **33** 4 p674

[10] Devillard, P. and Souillard, B. (1986),
"Polynomially Decaying Transmission for the Nonlinear Schrödinger Equation in a Random Medium" *Journal of Statistical Physics* **43** 3-4

[11] Dygas, M. M., Matkowsky, B. J., and Schuss, Z. (1987),
"Stochastic Stability of Nonlinear Oscillators," Preprint

[12] Flytzanis, C., (1984),
"Bistability, Instability, and Chaos in Passive Nonlinear Optical Systems", Lecture notes from the Proceedings of the Third International School on Condensed Matter Physics, <u>Nonlinear Phenomena in</u> <u>Solids–Modern Topics</u>, ed. M. Borissov, World Scientific

[13] Fürstenberg H. (1963),
"Noncommuting Random Products", *Trans. Am Soc.*, **108** p377-428

[14] Gibbs, H.M., McCall, S.M., Venkatesan, T.N.C., Gossard, A.C., Passner, A., Wiegmann, W. (1979),
"Optical Bistability in Semiconductors", *Appl. Phys. Lett.* **35** 6 p451

[15] Gibbs, H.M., Tarng, J.L., Weinberger, D.A., Tai, K. and Gossard, A.C., McCall, S.L., Passner, A., Wiegman, W. (1982),
"Room Temperature Excitonic Optical Bistability in a GaAs-GaAlAs Superlattice Etalon", *Appl. Phys. Lett.* 1 Aug, 1982

[16] Gihman, I.I., Skorohod, A.V. (1972),
<u>Stochastic Differential Equations</u> Springer Verlag

[17] Knapp, R. (1988),
"Nonlinearity and Localization in One Dimensional Random Media", PhD Thesis, New York University, October 1988

[19] Kohler, W. and Papanicolaou, G. (1973),
"Power Statistics for Wave Propagation in one Dimension and Comparison With Transport Theory", *J. Math. Phys.* **14**,p1733-1745 and **15** p2186-2197 (1974)

[18] Klyatskin, V.I. (1984),
"Solution of a Nonlinear Problem on Self-action of a Plane Wave in a Layered Medium Using the Immersion Method", Preprint

[20] Klyatskin, V.I., Kozlov, F.F., and Yaroshchuk, E.V. (1982),
"Reflection Coefficient in the One-dimensional Problem of Self-action of a Wave", *Sov. Phys. JETP* **55** 2 p220 American Institute of Physics

[21] Marburger, J. H. and Felber, F. S. (1978),
"Theory of a Lossless nonlinear Fabry-Perot Interferometer," *Physical Review A* **17** p335-342

[22] McCall, S.L., Gibbs, H.M., Venkatesan, T.N.C. (1975),
J. Opt. Soc. Am. **65** p1184

[23] Miller, D.A.B., Smith, S.D., and Johnston, A. (1979),
"Optical Bistability and Signal Amplification in a Semiconductor Crystal: Applications of New Low-power Nonlinear Effects in InSb", *Appl. Phys. Lett.* **35** 9 p658

[24] Papanicolaou, G. C. (1978),
"Asymptotic Analysis of Stochastic Equations," *M.A.A. Studies in Mathematics*, **18**, ed. M. Rosenblatt, M.A.A.

[25] Papanicolaou, G. C. (1977),
"Introduction to the Asymptotic Analysis of Stochastic Equations," *Lectures in Applied Mathematics* **16** p109-147

[26] Press, W., Flannery, B., Teukolsky S., Vetterling W. (1986),
<u>Numerical Recipes</u> Cambridge University Press

[28] Sheng, P., Zhang, B., White, B., Papanicolaou, G. (1986),
"Multiple Scattering Through Localization Length Scales", *Phys. Rev. Letters* **57**8 p1000-1003

[27] Sheng, P., Zhang, B., White, B., Papanicolaou, G. (1986),
"Minimum Wave Localization Length in a one Dimensional Random Medium", *Phys. Rev. B* **34** 7 p4757-4761

[29] Spigler, Renato (1986),
"Mean Power Reflection from a One-dimensional Nonlinear Random Medium", *J. Math. Phys.* **27** 7 p1760-1771

[30] Spigler, Renato (1985),
"Nonlinear Parametric Oscillations in Certain Stochastic Systems: A Random van der Pol Oscillator", *J. Stat. Phys.* **41** p175

[31] Warren, M.E., Koch, S.W., Gibbs, H.M. (1987),
"Optical Bistability, Logic Gating, and Waveguide Operation in Semiconductor Etalons", *IEEE Computer* **20** 12 p68-81

Wave Transmission in a One-Dimensional Nonlinear Lattice: Multistability and Noise

Yi Wan and C.M. Soukoulis

Ames Laboratory, USDOE and Department of Physics,
Iowa State University, Ames, IA 50011, USA

We studied a nonlinear Schrödinger equation of the tight binding form and the related problem of wave transmission in one-dimensional nonlinear lattice. By introducing a two-dimensional nonlinear mapping and applying the concepts and methods of nonlinear dynamics to the investigation of such mapping, we discuss some of the interesting properties of the transmission problem. The Holstein molecular crystal model is used as a physical system that may be relevant to the nonlinear transmission phenomenon described.

1. INTRODUCTION

The nonlinear Schrödinger equation is very useful in describing many physical systems involving nonlinearities, especially the propagation of waves in the continuous dispersive nonlinear media [1]. The success, in this respect , is due largely to the fact that in many cases for wave fields with small amplitudes and slow spatial variations, the nonlinear evolution equations for the fields can be approximated by nonlinear Schrödinger equations and various perturbation methods can be applied. However, in physical systems showing crystalline structure, the existence of a fundamental periodicity in certain circumstances makes the continuous medium approximation unnatural. Also, it is not clear what the physical consequences would be in a situation where the interaction between the effects of nonlinearity and periodic modulation is dominant. On the other hand, it is well-known in the field of nonlinear dynamical systems that many nonlinear systems represented by Hamiltonians of relatively simple form can exhibit rich and complex behaviors [2-4]. Such a class of nonlinear dynamical systems can be modelled by area-preserving nonlinear mappings on the Poincaré surface section. It has been shown that the complexity in their dynamical properties has its origin in the non-integrability of these nonlinear mappings [5,6]. Given such circumstances, it is interesting for us to see whether nonlinearity in certain simple physical systems, which can be related to some nonlinear dynamical models, would give rise to new physical phenomena. In this paper, we consider a one-dimensional lattice chain described by a nonlinear Schrödinger equation of the tight-binding form, and employ the concepts and methods of nonlinear dynamical mappings to study the problem of wave propagation in such a nonlinear system. Specifically, we consider a nonlinear sample of finite length embedded in an infinitely long linear lattice chain and a wave travelling in the linear part of the system arriving at one end of the nonlinear sample (Fig. 1). Under certain conditions, the incident wave may excite collective oscillations in the nonlinear medium. Such responses may be able to propagate through the nonlinear sample and emerge as a transmitted wave. If such a wave propagation persists in a stable state, the resulting steady wave field in the nonlinear medium can be described by a time-independent nonlinear Schrödinger equation. We considered a simple form of such an equation:

$$-\psi_{n-1} - \psi_{n+1} - \lambda|\psi_n|^2\psi_n = E\psi_n \ ,$$

where ψ_n denotes the wave function at the nth lattice site, E and λ are two real numbers, the energy of the wave and the nonlinear coupling constant, respectively. This form of the nonlinear Schrödinger equation was derived by Holstein in his model [7] for a diatomic molecular crystal chain where ψ_n represents the wave function of a single excess electron in the conduction band, E is the Fermi energy of the electron, and the nonlinear term, $\lambda|\psi_n|^2\psi_n$, originates from the short range electron-lattice interaction. The Holstein model

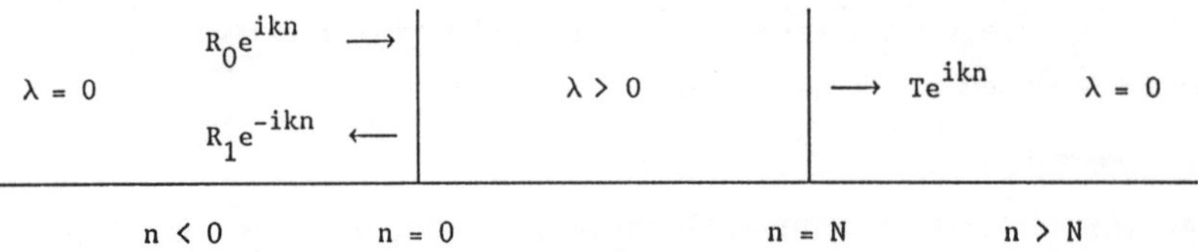

Fig. 1. The nonlinear wave transmission as defined by Eq. (1). The nonlinear medium ($\lambda > 0$) extends from $n = 0$ to $n = N$. There are incoming and reflected waves in the linear medium ($\lambda = 0$) to the left of the nonlinear sample ($N < 0$), whereas only the outgoing wave exists on the other side ($n > N$) of the nonlinear sample.

provides an example of the nonlinear transmission in electronic systems. We will take it as a typical physical system in our subsequent discussions. The Holstein Hamiltonian and its associated Schrödinger equation can also describe the motion of a single hole in the molecular crystal chain, provided λ is negative and E and k referring to the energy and wave vector of the hole, respectively. Thus, we will restrict our study to the electron case only, i.e., $\lambda > 0$.

The linear medium is described by the same tight-binding Schrödinger equation with $\lambda = 0$. The wave function in the linear sample takes the form of a Bloch wave. Considering an incident wave coming from left towards the nonlinear sample extending over N lattice sites, we define our transmission problem as the following (see Fig. 1):

$$\begin{cases} -\psi_{n+1} - \psi_{n-1} - \lambda|\psi_n|^2\psi_n = E\psi_n & 0 \leq n \leq N \\ \psi_n = R_o e^{ikn} + R_1 e^{-ikn} & n \leq 0 \\ \psi_n = Te^{ikn} & n \geq N \end{cases} \qquad (1)$$

In Eq. (1), we require $|R_o|^2 = |R_1|^2 + T^2$ to satisfy the conservation of probability current and choose the overall constant phase for the wave functions so that T is real. We study the relation between the intensities of the incident wave $|R_o|^2$ and the outgoing wave, T^2, i.e., the nonlinear response, and also their dependence on the parameters E, k, and λ. In the linear transmission problem (i.e., $\lambda = 0$), the wave function ψ_n for $0 < n \leq N$ has also the form of a Bloch wave described by a wave vector k related to E by $E = -2\cos k$. The transmission is practically possible only for E in the energy band ranging from -2 to 2. Taking into consideration the possible wave scatterings at the interfaces located at $n = 0$ and $n = N$, the properties of the transmission coefficient t defined by $t = T^2/|R_o|^2$ as a function of the parameters are well understood. But with the presence of the nonlinearity, we need to reexamine the dependence of the wave field inside the nonlinear medium on the boundary conditions at $n = 0$ or at $n = N$. As will be shown later, the wave functions ψ_n, $0 < n < N$, can be uniquely determined by the output Te^{ikn}, $n \geq N$, rather than by the input R_o. Thus, our strategy in solving the nonlinear transmission problem is to solve for the wave functions inside the nonlinear sample with given information on T, E, k, and N, and then determine R_o as a function of these quantities. It is clear that the methods commonly used in the linear transmission problem, such as that employing transfer matrices, are inadequate for our purpose. The complexity involved in our problem can be appreciated by looking at the "phase diagram" on the $(E, \lambda T^2)$ plane (see Fig. 2 and discussions in Section 3), where the region of $(E, \lambda T^2)$ corresponding to the bounded wave field in the nonlinear sample (shaded) is separated from that of diverging waves (i.e., infinite input!) by a sharp and complexly-shaped boundary. Such a "phase diagram" in the range $-2 < E < 2$ was first discovered by Delyon and co-workers [8] in their study of the nonlinear transmissions. By introducing a nonlinear dynamical mapping for the complex wave functions, we are able to study the "phase diagram" and the structures present in the transmission coefficient t. Moreover, we found chaotic behavior of the wave field (as a direct consequence of the non-integrability of the nonlinear mapping), contrary to the assertions of the previous authors [8,9]. In Section 2, we derive, for the transmission problem, a

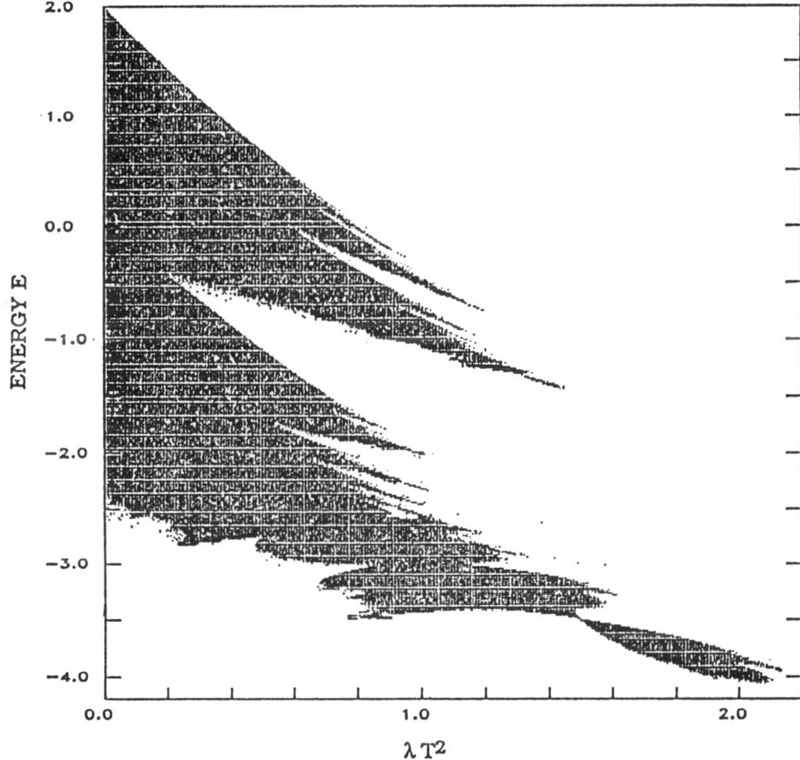

Fig. 2. The energy E versus the nonlinear strength λT^2. Transmitting (hatched) and non-transmitting (clear) regimes for the discrete nonlinear equation.

nonlinear mapping on the plane and investigate it in the context of nonlinear dynamics. The results of the nonlinear transmission problem defined in Eq. (1) are presented and discussed in Section 3.

2. NONLINEAR DYNAMICAL MAPPING ON THE PLANE

The nonlinear Schrödinger equation in Eq. (1) can be written as two equations containing real and imaginary parts of the wave function ψ_n:

$$\begin{cases} -x_{n+1} - x_{n-1} - \lambda(x_n^2 + y_n^2)x_n = Ex_n \\ -y_{n+1} - y_{n-1} - \lambda(x_n^2 + y_n^2)y_n = Ey_n \end{cases}, \tag{2}$$

where x_n and y_n are the real and imaginary parts of ψ_n. Equation (2) has the form of two coupled difference equations of the second order and represents a nonlinear mapping in terms of $(x_{n-1}, x_n, y_{n-1}, y_n)$. Therefore, the system operates in a four-dimensional space. A particular problem that may arise for such a nonlinear system involves the so-called Arnold's diffusion [3,4] which is known to exist, in general, in nonlinear dynamical mappings with degrees of freedom greater than two. As a consequence, one cannot guarantee the existence of a particular bounded orbit (x_n, y_n) for an arbitrarily long lattice. However, it can be shown that because of the conservation of probability current required in our problem, this four-dimensional mapping can be reduced to two-dimensional and the possibility of the Arnold's diffusion can be ruled out.

In order to derive the reduced mapping we write the wave function and Eq. (2) in polar coordinates: $\psi_n = r_n e^{i\theta_n}$ and

$$\begin{cases} r_{n+1}\cos\Delta\theta_{n+1} + r_{n-1}\cos\Delta\theta_n = 2f(r_n) \\ r_{n+1}\sin\Delta\theta_{n+1} - r_{n-1}\sin\Delta\theta_n = 0 \end{cases} \quad , \tag{3}$$

where $\Delta\theta_n = \theta_n - \theta_{n-1}$ and $f(r) \equiv -\frac{1}{2}r(\lambda r^2 + E)$. The second equation defines an integral of motion for Eq. (3):

$$J \equiv r_n r_{n-1}\sin\Delta\theta_n \quad , \tag{4}$$

which has the physical meaning of probability current. A similar situation arises in a model for the motion of the colliding proton beams in a storage ring [10]. By introducing new variables as Bountis and co-workers did in their study [10]:

$$\begin{cases} u_n \equiv r_n^2 \\ v_n \equiv J\cot\Delta\theta_n \end{cases} \quad , \tag{5}$$

we obtain from Eq. (4) and the first equation in Eq. (3) the reduced mapping S:

$$S : \begin{cases} u_{n-1} = \frac{1}{u_n}(v_n^2 + J^2) \\ v_{n-1} = -v_n - u_{n-1}(u_{n-1} + E) \end{cases} \quad . \tag{6}$$

In the mapping S, the variables u and v have been scaled by the nonlinear coupling constant $\lambda : (u, v) \to (\lambda u, \lambda v)$, and λ is absorbed into $J : J \to \lambda J$. Thus, by redefining the variables, we have reduced the number of parameters in mapping S from three to two: the energy E, and the current J which describes the strength of the nonlinearity and is determined from λ and the output of the transmission problem. The mapping S is understood in the following way. From the wave functions at the boundary of the nonlinear sample, ψ_N and ψ_{N+1} in our case, we obtain immediately r_N, r_{N+1}, θ_N, and θ_{N+1}. Hence, u_N, u_{N+1}, v_{N+1}, together with λ, we determine J. From the second equation in Eq. (6), we obtain v_N and then use (u_N, v_N) to initiate the iteration S. On the other hand, having obtained a series of (u_n, v_n) we immediately have r_n and $\Delta\theta_n$ in $[0, \pi]$ if $J > 0$ (or in $[\pi, 2\pi]$ if $J < 0$), thus a particular $\theta_m = -\sum_{n=m}^{N} \Delta\theta_{n+1}$. This enables us to determine the wave functions in the nonlinear medium in terms of the (bounded) orbits of mapping S on the (u, v) plane which is the Poincaré surface section of the corresponding nonlinear dynamical system (with n playing the role of the discrete time). For the special case $J = 0$ (e.g., $k = 0$ or π, discussed later), we use directly Eq. (2) which is now decoupled and becomes the nonlinear Schrödinger equation in the real domain that we discussed elsewhere [11].

Before going into numerical calculations, it is important first to study some general properties of the mapping S. In fact, it proves these properties are essential in understanding our problem of transmission in nonlinear medium. First, the mapping S preserves locally the weighted measure $(\frac{1}{u_n}du_n dv_n)$ since its local Jacobian $(det(DS^{(n)}) = \frac{u_{n-1}}{u_n})$ is not unity in general. $DS^{(n)}$ is the tangent mapping of S at (u_n, v_n):

$$\begin{pmatrix} \delta u_{n-1} \\ \delta v_{n-1} \end{pmatrix} = DS^{(n)} \begin{pmatrix} \delta u_n \\ \delta v_n \end{pmatrix}, \quad DS^{(n)} = \begin{pmatrix} -\frac{u_{n-1}}{u_n} & \frac{2v_n}{u_n} \\ \frac{u_{n-1}}{u_n}(2u_{n-1} + E) & -\frac{2v_n}{u_n}(2u_{n-1} + E) - 1 \end{pmatrix} . \tag{7}$$

However, the area is indeed preserved for a periodic orbit after mapping through the complete period q: $\prod_{n=1}^{q} det(DS^{(n)}) = \prod_{n=1}^{(q)} \frac{u_{n-1}}{u_n} = 1$. Thus, S is equivalent topologically to an area-preserving mapping. Moreover S is locally area-preserving if the coordinates $(\ell n u_n, v_n)$ are used instead. Consequently, we will plot all the orbits in the plane $(\ell n u_n, v_n)$. Second, we show that mapping S is reversible. In fact, if we write $S = S_2 \circ S_1$, we factorize S into involutions S_1 and S_2

$$S_1 : \begin{cases} u' = \frac{1}{u}(v^2 + J^2) \\ v' = v \end{cases} ; \quad S_2 : \begin{cases} u' = u \\ v' = -v - u(u + E) \end{cases} . \tag{8}$$

The involutions S_1 and S_2 satisfy $S_1^2 = S_2^2 = 1$ so that $S^{-1} = S_1 \circ S_2$. Such inversibility seems obvious in terms of the original nonlinear Schrödinger Eq. (1), since it generates the same wave functions (ψ_n) with given (ψ_0, ψ_1) or (ψ_N, ψ_{N+1}) in the series. Geometrically the mapping S_1 has the effect of inversion along the u-axis about the $(u > 0)$ branch of the hyperbola $u^2 - v^2 = J^2$, and S^2 represents a reflection about the parabola $v = -\frac{1}{2}u(u + E)$ along the v-axis. The two curves $u = \sqrt{v^2 + J^2}$ and $v = -\frac{1}{2}u(u + E)$ are the symmetry lines of S_1 and S_2, respectively, since each is actually formed by the set of fixed points of the corresponding mapping.

The orbits generated by the mapping S on the (u, v) plane can be either bounded or diverging. Only those from the former category contribute to wave transmissions in our problem. The bounded orbits are organized into a hierarchy of periodic orbits of different periods, as is generally the case for area-preserving nonlinear mappings. Each stable periodic orbit is surrounded by higher order periodic and quasi-periodic orbits, whereas around the unstable periodic orbits there are locally chaotic orbits. The stability of a periodic orbit of period q is determined by its residue R, defined as $R = \frac{1}{4}[2 - Trace(\prod_{n=1}^{q} DS^{(n)})]$. The periodic orbit is stable when $0 < R < 1$ whereas, it is unstable if $R < 0$ or $R > 1$. Notice that the periodic orbit and its residue are functions of the parameters in the mapping S. As an example, the period-1 periodic orbit is given by:

$$\begin{cases} u = -E - 2cosk \\ v = -(E + 2cosk)cosk \end{cases} , \tag{9}$$

with $R = 1 - \frac{E}{2}cosk - 2cos^2k$. We show in Fig. 3 the periodic orbit of period 5 and the surrounding quasi-periodic and chaotic orbits. The presense of the chaotic orbit is closely related to the non-integrable characteristics of the nonlinear dynamical mapping. Notice

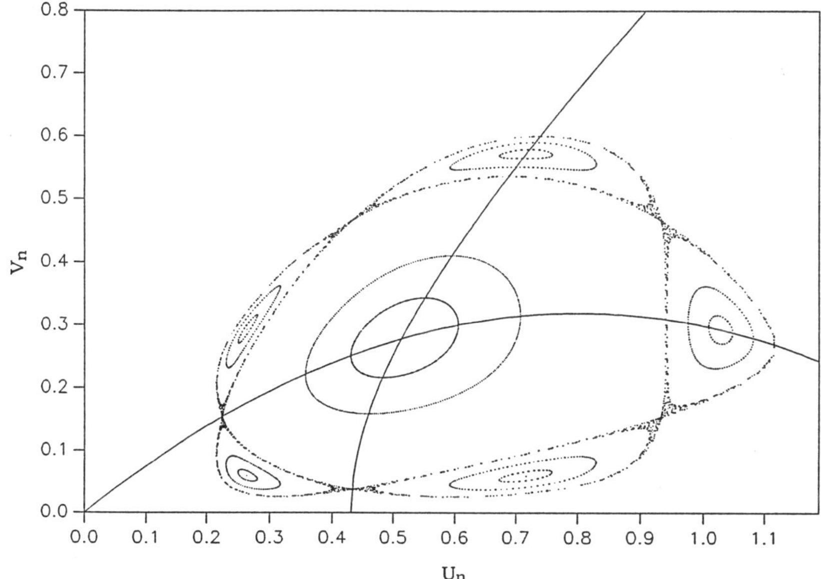

Fig. 3. The period-5 periodic orbit for $E = -1.598$ and the surrounding quasi-periodic and chaotic orbits on the (u, v) plane. The solid lines are part of the symmetry lines of mapping S_1 and S_2.

how the orbits are arranged about the two symmetry lines. In fact the symmetry lines make it much easier to locate numerically a periodic orbit. Since the starting point of the iterations (u_N, v_N) is always on the symmetry line of S_1, i.e., $u_N = \sqrt{v_N^2 + J^2}$ (c.f. Section 3), it is sufficient to do a one-dimensional search for a periodic orbit of a given period.

3. RESULTS AND DISCUSSIONS

Now we consider the orbits of mapping S that may contribute to the wave transmission in our problem. At the right end of the nonlinear sample, the wave functions representing the transmitted wave are simply the Bloch waves: $\psi_n = T e^{ikn}$, $n \geq N$. From Eqs. (4) and (5), we have $J = \lambda T^2 \sin k$, $u_N = u_{N+1} = \lambda T^2$ and $v_{N+1} = \lambda T^2 \cos k$ (we take $\lambda = 1$ in all the numerical calculations) so that (u_N, v_N) is on the symmetry line of S_1. In general, the energy band in a linear medium ($n > N$ and $n < 0$) may not range from -2 to 2, presumably due to the potential barriers across the interfaces of the linear and nonlinear media at $n = 0$ and $n = N$, where voltage differences may exist. In fact, it is necessary to have the lower edge of the energy band in the linear medium below -2 to observe the transmission phenomenon occurring at a Fermi level in the "energy gap" ($E < -2$ in this case) in the nonlinear sample. Such phenomenon is interesting because the wave transmission in this case is due entirely to the nonlinearity. Furthermore, when the Fermi level is deep in the "energy gap," a critical current J_c must be exceeded in order to send a wave through the nonlinear sample (see discussion in Section 3). Based on the above consideration, it seems that in the general case of the electronic transmission problem, we should determine the Fermi level E and the wave vector k independently. However, to avoid complexity and to focus on the basic nonlinear effect in wave transmission, we choose to simplify the problem by considering the following two cases:

(1) for $-2 \leq E \leq 2$ we adopt the relation $E = -2\cos k$;

(2) for $E < -2$, we fixed a value for the wave vector $k = \frac{2\pi}{100}$.

The "phase diagram" in Fig. 2 is plotted by choosing E, J, and (u_N, v_N) as discussed above and by iterating the mapping S. If the iterations give a bounded orbit (a cut-off is used in the numerical procedure), then the point $(\lambda T^2, E)$ falls in the shaded region which we call the "stable region," otherwise it is in the clear region. We found that the boundary separating these two regions is insensitive to either the cut-off or to the number of interactions as long as a moderate number of the order 10^2 is exceeded. The points $(\lambda T^2, E)$ corresponding to a particular stable periodic orbit form a curve extending from the vertical axis to a point on the boundary of the shaded region where the orbit loses its stability. The surrounding quasi-periodic orbits have their $(\lambda T^2, E)$ points around the curve and together they produce the branch-like structure on the "phase diagram."

Having discussed the general behavior of the wave function in the nonlinear sample, we now turn to the transmission problem introduced in Section 1. The quantity of interest, $|R_o|^2$ is numerically calculated by iterating the mapping S with a specified E and k for various values of $T(\lambda = 1)$. The calculations are performed for two typical cases, one for $k = \frac{\pi}{3}$ and $E = -1$, the other for $k = \frac{2\pi}{100}$ and $E = -2.01$ (in the energy "gap"), both with sample length $N = 200$. The results and the interesting features of the nonlinear transmission are summarized in Figs. 4 and 5. Figures 4(a) and (b) show, in different scales, the intensity of the incident wave $|R_o|^2$ is a single-valued function of the transmitted intensity T^2. On the other hand, T^2 has a multi-valued functional dependence on $|R_o|^2$. Such a dependence is the source of bistability, or multistability, in general. Figure 4(c) presents the same information in terms of the transmission coefficient t as a function of T^2. Close examination of the "curves" reveals the existence of the distinctive features. There are always "flat" or monotonic regions on the curve separated by "oscillatory" regions.

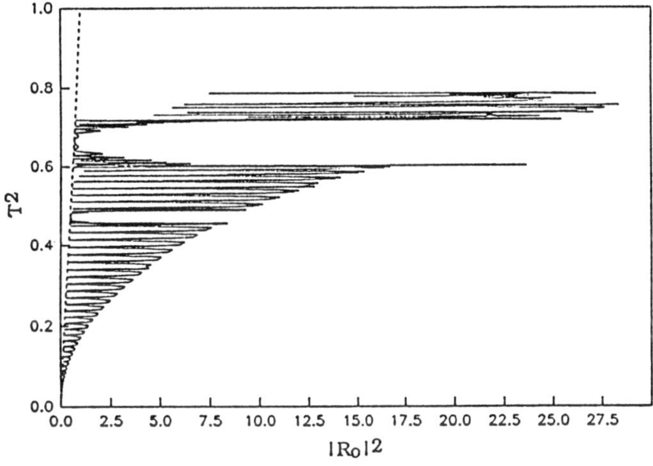

Fig. 4(a).

The transmitted intensity T^2 vs the incident intensity $|R_o|^2$ for $E = -2.01$. The dashed line corresponds to $T^2 = |R_o|^2$ which describes the linear transmission.

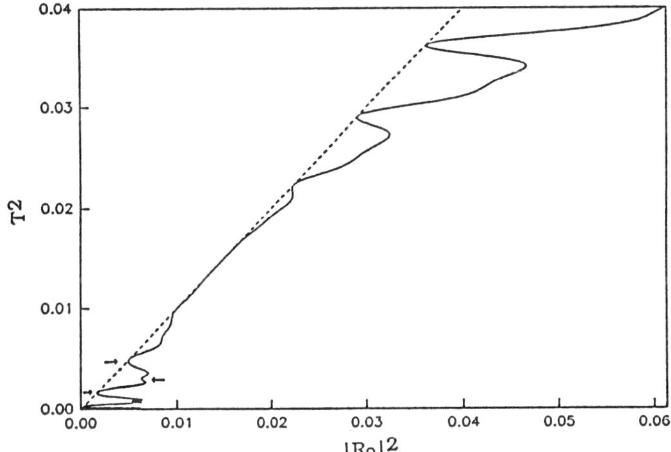

Fig. 4(b).

Enlargement of the section around the origin of Fig. 4(a).

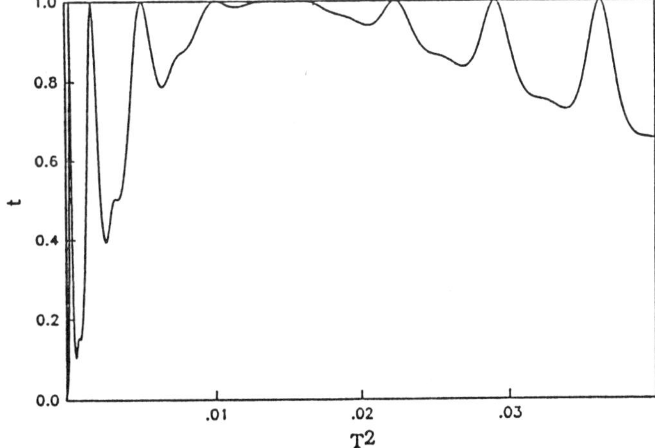

Fig. 4(c).

The transmission coefficient t vs T^2 for $E = -2.01$. T^2 has the same range as in Fig. 4(b).

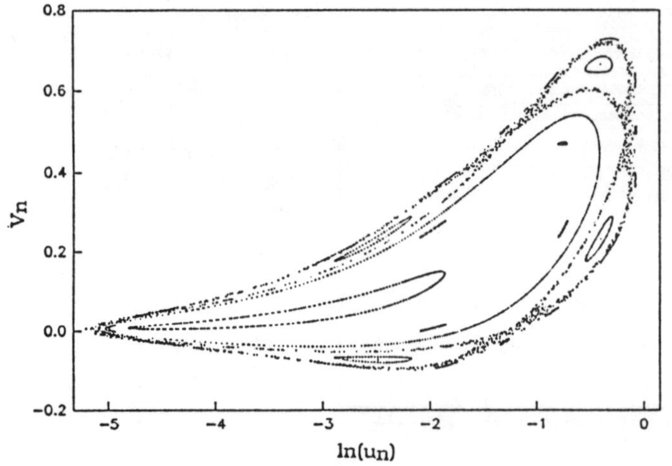

Fig. 4(d).

Some of the orbits corresponding to values of T^2 in Fig. 4(a) on the $(\ell nu, v)$ plane.

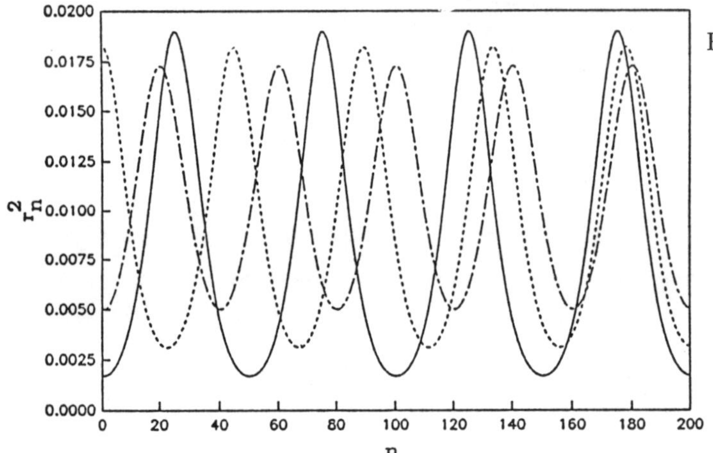

Fig. 4(e).

The wave amplitude r_n^2 vs the site n for the nonlinear segment for $E = -2.01$ and $T^2 = 1.7 \times 10^{-3}$ (———), 3.1×10^{-3} (— — — —) and 5.0×10^{-3} (· — · — ·). The T^2 values are also indicated by arrows in Fig. 4(b).

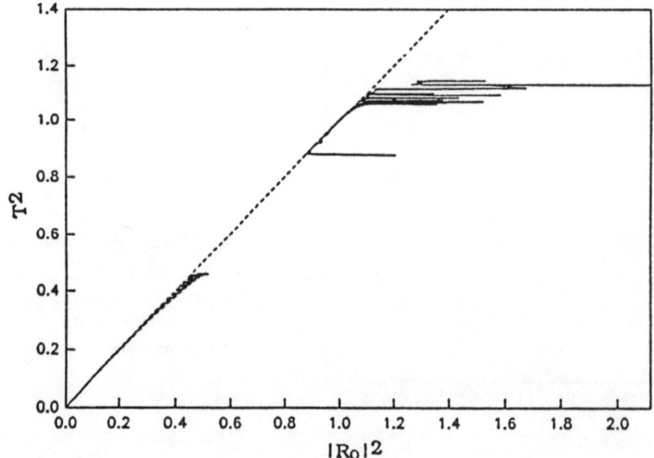

Fig. 5(a).

The transmitted intensity T^2 vs the incident intensity $|R_o|^2$ for $E = -1.0$. The transmission gap between $T^2 = 0.45$ to 0.92 is clearly seen.

34

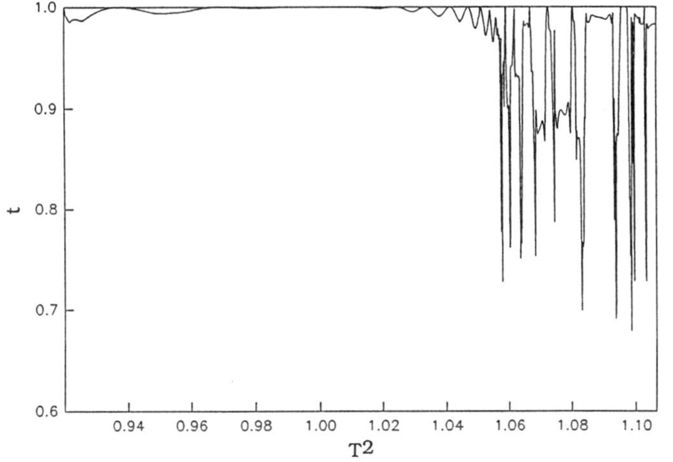

Fig. 5(b). The transmission coefficient t vs T^2 for $E = -1.0$.

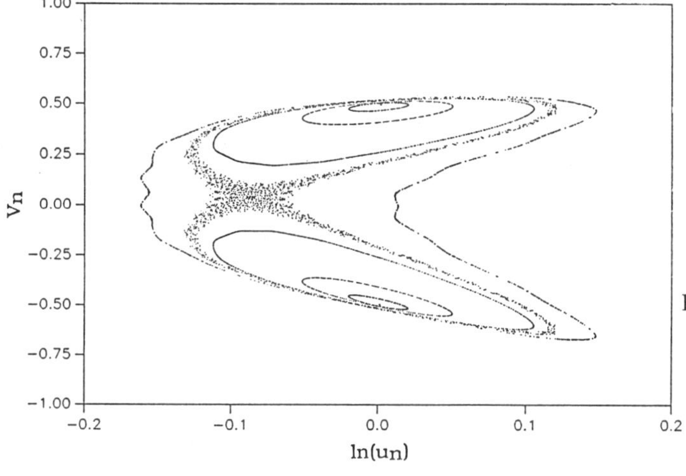

Fig. 5(c). Some of the orbits corresponding to values of T^2 in Fig. 5(b) on the $(\ell n u, v)$ plane.

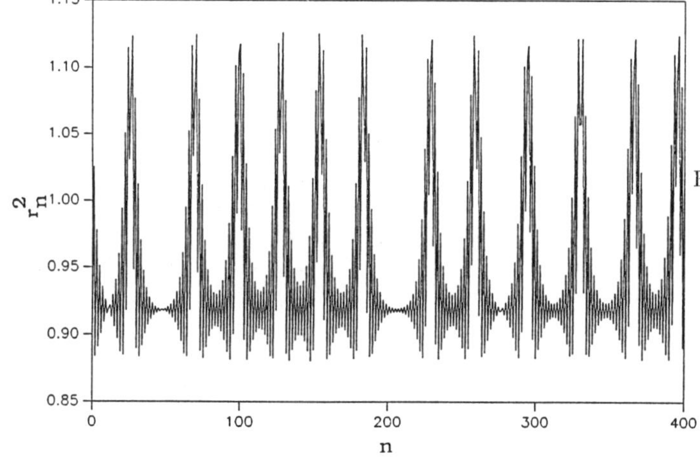

Fig. 5(d). The wave amplitude r_n^2 vs n for the nonlinear segment for $E = -1.0$ and $T^2 = 1.0793$, which corresponds to the chaotic orbit around the unstable period-2 orbit in Fig. 5(c).

Most parts of the regions show oscillations with a roughly uniform frequency. However, there are bursts of violent "oscillations" immediately around the flat regions, especially when T is large. Eventually the curve becomes irregular and disappears at the edges of the gaps on the $|R_o|^2$ vs T^2 curve, inside which no transmission is possible. Such features can be organized in reference to the "phase diagrams" in Fig. 2, where for a given value of E, T^2 can be either in the shaded region or in the clear region where no wave is transmitted. The tendency is that the regions on the curve corresponding to T^2 in the middle parts of the stable region are relatively smooth and as T^2 gets closer to the boundaries of the stable region, the corresponding curves become oscillatory and irregular. Figure 4(d) establishes a connection between the $|R_o|^2$ vs T^2 curve and the orbits on the $(\ell nu, v)$ plane, which can help us to understand the features described above. The orbits plotted in Fig. 4(d) are generated from some typical values of T^2 on the $|R_o|^2$ vs T^2 curve. Specifically, we plot several periodic orbits of lower periods and the quasi-periodic orbits around them. The inner most orbit in Fig. 4(d) is one of the quasi-periodic orbits surrounding the period-1 periodic orbit at $T^2 = 0.0139$ (c.f. Eq. (9) and Fig. 4(b)). Such orbits are responsible for the part of the $|R_o|^2$ vs T^2 curve below the point $T^2 \approx 0.45$ in Fig. 4(a). As T^2 increases, the orbits of larger size and with higher periods come into play, contributing to the transmission. The flat regions around $T^2 = 0.47$ and $T^2 = 0.71$ correspond to the orbits surrounding the period-5 and period-4 periodic orbits, respectively, which we refer to as period-5 and period-4 "satellites." The chaotic orbits which originated from the unstable period-4 periodic orbits are evident in Fig. 4(d) and give rise to the irregular "bursts" around the region at $T^2 = 0.71$. Similar chaotic orbits from the the unstable period-5 periodic orbit occupy a very narrow area on the $(\ell nu, v)$ plane and are not shown on the plot. Notice that there are still quasi-periodic orbits similar to the inner most orbit surrounding the period-5 satellites. In fact, such quasi-periodic orbits exist up to the edges of the gaps on the $|R_o|^2$ vs T^2 curve, and it is such orbits that keep the local chaotic orbits from escaping to infinity. In Fig. 4(c) we plot the wave intensity as a function of the position n to illustrate the relation between the $|R_o|^2$ vs T^2 curve. All three waves plotted correspond to the quasi-periodic orbits around the period-1 orbit at $T^2 = 0.0139$ and their corresponding T^2-values are indicated in Fig. 4(b). Since the waves have values of T^2 smaller than 0.0139 and the iterations all begin with $r_N^2 = r_{N+1}^2 = T^2$, the waves have intensities at their minima at $n = N = 200$. However, at the left end of the nonlinear sample, the intensity of the reflected wave is given by

$$|R_1|^2 = \frac{1}{4 sin^2 k}[(u_o - u_N) + (u_1 - u_{N+1}) - 2cosk(v_1 - v_{N+1})] \geq 0 \ .$$

Thus, only the waves that have an integral number of complete resonances inside the nonlinear sample can cause $|R_1|^2 = 0$, i.e., producing resonant transmission with $t = 1$. It is also evident that as the sample length N is varied while the remaining parameters are unchanged, the condition for resonant transmission will be changed. This results in shifts of the resonant transmission peaks on the t vs T^2 plot.

Now we turn to Fig. 5 with E in the "energy band." Comparing the plots in Fig. 5 with the corresponding plots in Fig. 4, we see that the two cases share the properties discussed above. However, an important discrepancy appears at the small T^2 segments of their $|R_o|^2$ vs T^2 curves (note the oscillations as $T^2 \to 0$ in Fig. 4(b), whereas they are absent in Fig. 5(a)). The reason for this discrepancy is that we adopt the relation $E = -2cosk$ only for $-2 < E < 2$ so that the period-1 orbit is located at finite u for $E < -2$ (c.f. Eq. (9)). Hence, reducing T^2 from only the period-1 periodic orbit has a similar effect as increasing T away from it. In fact, we enter the truely nonperturbative regime when E is considerably below -2 such that the shaded region moves away from $T = 0$ (when $E \leq -2.45$ in Fig. 2), consequently a critical value of T or J must be exceeded to make transmission possible. In other words, the linear transmission is unstable when E is in the energy gap. Finally, in Fig. 5(d), we plot r_n^2 vs n for the chaotic orbit around the unstable period-2 periodic orbit (c.f. Fig. 5(c)). Such a chaotic wave field will certainly create noisy responses in the transmission process, even for an arbitrarily small change in the values of the parameters in the system, such as E or k as well as the nonlinear sample length N.

36

In conclusion, the properties we discussed for the simple nonlinear transmission problem should be of general use in the nonlinear systems that can be related to certain type of nonlinear dynamical mappings. As we have demonstrated, these properties are the direct consequences of theorems commonly shared by the entire class of area-preserving nonlinear mappings. It would be very interesting and beneficial to study such relations in more realistic physical systems and in experimental conditions.

ACKNOWLEDGMENTS

Ames Laboratory is operated for the U.S. Department of Energy by Iowa State University under contract no. W-7405-ENG-82. This investigation was supported by the Director for Energy Research, Office of Basic Energy Sciences.

REFERENCES

1. R. K. Dood, J. C. Eilbeck, J. D. Gibbon, and H. C. Morris: *Soliton and Nonlinear Wave Equations* (Academic Press, Inc., London, 1984).

2. A. J. Lichtenberg and M. A. Lieberman: *Regular and Stochastic Motion* (Springer-Verlag, New York, Inc. 1983).

3. V. I. Arnold and A. Avez: *Ergodic Problems of Classical Mechanics* (W. A. Benjamin, Inc., 1968).

4. M. V. Berry: In *Topics in Nonlinear Dynamics*, ed. by S. Jorna, American Physical Society Conference Proceedings Series Vol. **46** (AIP, New York, 1978); B. V. Chirikov: *Phys. Rep.* **52**, 263, (1979).

5. J. M. Greene: *J. Math. Phys.* **20**, 6, 1183 (1979); J. M. Greene, R. S. Mackay, F. Vivaldi, and M. J. Feigenbaum: *Physica* **D3**, 468 (1981).

6. R. H. G. Helleman: In *Universality in Chaos*, ed by P. Critanović (Adam Hilger Ltd., Bristol, 1984).

7. T. D. Holstein: *Ann. Phys.* (New York) **8**, 325 (1959); L. A. Turkevich and T. D. Holstein: *Phys. Rev.* **B35**, 7474 (1987).

8. F. Delyon, Y. E. Levy, and B. Souilard: *Phys. Rev. Lett.* **57**, 2010 (1986).

9. L. Kahn, N. S. Almeida, and D. L. Mills: *Phys. Rev.* **B37**, 8072 (1988).

10. T. C. Bountis, C. R. Emmhizer, and R. H. G. Helleman: In *Long-Time Prediction in Dynamics*, ed. by C. W. Horton, Jr., L. E. Reichl and V. G. Szebehely (John Wiley & Sons, Inc., 1983).

11. Yi Wan and C. M. Soukoulis: to be published.

A New Look at Hopping, Trapping and Anderson Localisation

P. Phillips[1], K. Kundu[1], D.H. Dunlap[1], and P.E. Parris[2]

[1]Department of Chemistry, Massachusetts Institute of Technology,
  Cambridge, MA 02139, USA
[2]Department of Physics, University of Missouri, Rolla, MO 65401, USA

Abstract: We review a new set of transformations that provide a direct route to
the diffusion constant in the random hopping and trapping models for transport
among a random distribution of localised electronic states. We then show how
these techniques can be used to study the Anderson transition starting from a
generalised master equation with a nearest neighbour memory function.

1. Introduction

In this paper we review some new techniques we recently developed for studying the
transport of electrons among a random distribution of localised electronic states
[1-4]. We then show how these techniques can be used to study the Anderson transition [5]. The starting point for our work on the Anderson transition is the generalized master equation (GME) with an approximate memory function. Our analysis
shows that a GME with a nearest-neighbour memory function predicts accurately the
Anderson transition on a Cayley tree of connectivity K [5].

2. Hopping and Trapping Transport: an Overview

The two basic models that have been advanced to describe the incoherent conduction
of electrons among a random distribution of localised electronic states are the
random trapping and hopping models [6]. In the random trapping (RT) model, an
electron performs a series of incoherent, uncorrelated nearest-neighbour jumps on
an otherwise ordered array of random symmetric wells [6]. One symmetric well is
placed on each lattice site. The depth of the well on each lattice site is a
random variable determined by a single probability distribution. Because the
wells are symmetric, the jump rate at site n, ω_n to any of its nearest neighbours
is determined solely by the depth of the well at site n. It is for this reason
that the RT model is referred to as a site disordered model. In contrast, the
random hopping (RH) model is a bond-disordered model in which the localised electronic states are separated by symmetric barriers of varying height [6]. Randomness in the barrier heights gives rise to a random distribution of symmetric hopping rates. Here again the hopping rates are independent stochastic variables
determined by a single probability distribution.

The fundamental difference between these models is that the RT model does not
sustain diffusive transport in any spatial dimension anytime a finite fraction of
the hopping rates are identically zero [6]. This result can easily be understood
from the equilibrium site probabilities, P_n^{eq} in both models. In the RT model,
$P_n^{eq} \propto \omega_n^{-1}$, whereas $P_n^{eq} \propto N^{-1}$ (N the number of sites) in the RH model [6]. Consequently, a particle placed at some initial site in the RH model explores all sites
to which the initial site is connected with equal probability. On the other hand,
a particle in the RT model at long times explores sites with probability ω_n^{-1}. As
a result, the particle spends the most time in the deepest traps and never leaves
a trap if it is infinitely deep. That is, in the RT model at long times all the
particles collect into the sites for which $\omega_{n=0}$ (the infinitely deep traps). In
essence, the RT model unlike the RH model does not possess a dynamical percolation
transition. Such a transition occurs in the RH model for d>1 whenever the distri-

Springer Proceedings in Physics, Vol. 39   **Disorder and Nonlinearity**
Editor: A.R. Bishop   © Springer-Verlag Berlin, Heidelberg 1989

bution function contains some finite fraction of vanishing hopping rates. In fact, the RH model is the standard model for classical bond percolation and is generally applicable to physical systems in which a continuous path must form among the conduction sites for transport to obtain, for example impurity conduction in a semiconductor. The RT model is used in the context of anomalous diffusion [6] and transient photocurrent experiments [7].

As a result of the physical relevance of the RT and RH models, much theoretical effort has been expended to calculate their transport properties [8-12]. Let $P_n(t)$ be the probability that the electron is on the nth site. The discrete Fourier transform

$$\tilde{P}(q,t) = \sum_n e^{iqn} P_n(t) \tag{1}$$

which defines the probability Green function has served as the starting point for all of these investigations, primarily because the second q-derivative of (1) is related to the mean-square displacement. While the standard approach [8-12] based on (1) has provided much insight into the transport properties, we have found that a new set of transformations provide a much more direct and powerful route to the diffusion coefficient.

3. New Transformations for Hopping and Trapping Transport

To illustrate our approach, consider the equations of motion for the site probabilities

$$\dot{P}_n(t) = \begin{cases} \omega_{n+1}P_{n+1}(t) + \omega_{n-1}P_{n-1}(t) - 2\omega_n P_n(t) & \text{(RT)} \tag{2} \\[2mm] \omega_{n+1}(P_{n+1}(t) - P_n(t)) + \omega_n(P_{n-1}(t) - P_n(t)) & \text{(RH)} \tag{3} \end{cases}$$

in the 1-dimensional RT and RH models where ω_n is the hopping rate from site n to site n-1. The symmetric well constraint in the RT model requires that $\omega_{n \to n-1} = \omega_{n \to n+1}$, whereas the bond symmetry in RH requires $\omega_{n+1 \to n} = \omega_{n \to n+1}$. For the localised initial condition, $P_n(t=0) = \delta_{n,0}$, the diffusion coefficient [1,2,11]

$$D(t) = \ell^2 \langle \sum_n n^2 \dot{P}_n(t) \rangle = \begin{cases} 2\langle \sum_n \omega_n P_n(t) \rangle & \text{(RT)} \tag{4} \\[2mm] \langle \sum_n \omega_n(P_{n+1} - P_n) \rangle + 2\langle \sum_n n\omega_n(P_{n+1} - P_n) \rangle & \text{(RH)} \tag{5} \end{cases}$$

can be written succinctly in terms of the site exit probability $I_n(t) = \omega_n P_n(t)$ in the RT model and the probability current $J_n(t) = \omega_n(P_{n+1}-P_n)$ in the RH model. In eqs. (4) and (5), the angle brackets $\langle \cdots \rangle$ signify an average over the distribution of hopping rates. Eqs. (4) and (5) suggest that rather than eq. (1), we should focus on the new generating functions [1,2]

$$p^{RT}(q,t) = \sum_n e^{iqn} I_n(t) \tag{6}$$

and

$$p^{RH}(q,t) = \sum_n e^{iqn} J_n(t) \tag{7}$$

For both the RT and RH models, these generating functions provide a direct route to the diffusion coefficient:

$$D(q,t) = \begin{cases} \langle p^{RT}(q=0,t) \rangle & \tag{8} \\[2mm] \langle p^{RH}(q=0,t) \rangle - 2i \frac{\partial}{\partial q} \langle p^{RH}(q,t) \rangle \Big|_{q=0} & \tag{9} \end{cases}$$

Hence, once $P^{RT}(q,t)$ and $P^{RH}(q,t)$ are calculated, the transport properties follow immediately without the complication of the two q-derivatives of the generating function as in the standard approach.

Of course it would be fruitless to define the new generating functions, eqs. (6) and (7), if they could not be calculated simply. We have shown, however, that unlike the standard approach, exact solvable integral equations can be obtained for $P^{RT}(q,t)$ and $P^{RH}(q,t)$ directly from the equations of motion in any spatial dimension [1-4]. Consider, for example, the exit probability generating function in the RT model. We have already pointed out that the q=0 limit of $P^{RT}(q,t)$ defines the diffusion coefficient. It is appropriate to expand the frequency-dependent diffusion coefficient in terms of the inverse moment $z_n = <[1/\omega-<1/\omega>]^n>$ because the zero-frequency diffusion constant in the RT model irrespective of the spatial dimension is $D_0 = <1/\omega>^{-1}$ [1,8]. The generating function $P^{RT}(q,t)$ leads naturally to the inverse moment expansion because the closure relation [1] for the site probabilities

$$P_n(t) = \frac{1}{\omega_n} \int_{-\pi}^{\pi} P^{RT}(q,t)e^{-iqn} \frac{dq}{2\pi} \tag{10}$$

is of the form $\omega_n^{-1} \times$ (some function). The inverse moment expansion is easily obtained by 1) multiplying the equations of motion by e^{iqn}, 2) summing over n, and 3) adding and subtracting $<\omega^{-1}> \times$ (some function). In d-dimensions, the diffusion coefficient for a localised initial condition is of the form

$$D(\varepsilon)\lim_{\varepsilon\to 0} \sim \frac{D_0\ell^2}{2}[1 + z_2c_d\varepsilon + \begin{cases} \tilde{c}_d \ \varepsilon^{d/2} & d \text{ odd} \\ \tilde{c}_d \ \varepsilon^{d/2} \ \ell n\varepsilon & d \text{ even} \end{cases}] \tag{11}$$

where $\tilde{c}_d$ and c_d are constants determined by the spatial dimension [1]. Eq. (11) is in agreement with the results from the standard approach when localised initial conditions are used [13].

Bond-disordered systems are considerably more complicated because the inverse moment expansion is appropriate only for d=1 [2,9,10,14]. Generalising the flux method to higher dimensions involves two distinct steps. First, a flux generating function of the form of $P^{RH}(q,t)$ must be introduced along each direction of the crystal. The appropriate quantity is [2,3]

$$I^{\alpha}(\underset{\sim}{q},\varepsilon) = \sum_{\underset{\sim}{n}} \int_0^{\infty} e^{-\varepsilon t} \ \omega_{\underset{\sim}{n}}^{\alpha}(P_{\underset{\sim}{n}}-P_{\underset{\sim}{n}-\hat{\alpha}})e^{-i\underset{\sim}{q}\cdot\underset{\sim}{n}} \tag{12}$$

where α is a unit vector pointing along the α-direction from site $n = (n_1,n_2,\cdots n_d)$ to its nearest neighbour along that direction. In eq. (12) ε is the Laplace variable conjugate to time. $I^{\alpha}(q,t)$ is related to the diffusion constant, $D^{\alpha\alpha}$, along the α-direction through eq. 9. For an isotropic system $D_{\alpha\alpha}$ is independent of direction. Second, d-coupled integral equations must be constructed and solved for the I^{α}'s. The integral equations can be constructed directly from the equations of motion for P_n and $P_{n-\alpha}$ once a uniform system and a perturbation V are defined. The resultant equations of motion for the I^{α}'s can be written succintly as $[\tilde{s}+H - \varepsilon V]I = i(0)$ or $G^{-1}I = i(0)$ where $i(0)$ and I are column vectors with components $(1-e^{-iq_{\alpha}})$ and $I^{\alpha}(q,\varepsilon)$, respectively, $H_q^{\mu\alpha} = (1-e^{iq_{\mu}})(1-e^{-iq_{\alpha}})$ and the perturbation $\tilde{V}_{nm}^{\alpha\mu}=(c^{-1}-1/\omega_n^{\alpha})\delta_{\alpha\mu}\delta nm$. The parameter c is the uniform system hopping rate and $\tilde{s} = \varepsilon/c$. Note that the fluctuation V is entirely site-diagonal and would be site off-diagonal had we chosen to work with the probabilities as opposed to the fluxes. Because V is site-diagonal standard, t-matrix expansions [4] can be used to calculate the Green function, G, whose diagonal (in q-space) elements are

related to the diffusion coefficient through $\lim_{q \to 0} \langle G \rangle \alpha \alpha = D(\varepsilon)\delta(q')$. We [3] have
used the t-matrix expansion of G to calculate the zero-frequency diffusion con-
stant for the d=3 bond percolation problem. For a percolation distribution of the
form $\rho(\omega_n) = p\delta(\omega_n-\omega) + (1-p)\delta(\omega_n)$, setting $c=\omega$ results in an expansion for $D(\varepsilon=0)$
in powers of the fraction of broken bonds, (1-p). To order $(1-p)^3$, we obtain a
percolation threshold of P_c= 0.252 in excellent agreement with the exact value of
0.249 [3]. The conductivity exponent was t=1.2 as opposed to the accepted value
of t=1.6±0.1. In d=2, the exact value of p_c=0.5 is predicted. Hence, unlike the
standard approach, the flux transformations lead to accurate results for bond
percolation problems at low order in perturbation theory.

4. Anderson Localisation from a Generalised Master Equation

We now consider the transport of a single electron on an energetically-disordered
lattice in the tight-binding approximation. The nearest-neighbour matrix element
is V and ε_n is the random energy at site n. To apply the techniques just presented to the localisation-delocalisation transition in this model [14], it is necessary to construct a quantum mechanical equation for the time evolution of the site
probabilities. For a localised initial condition, the appropriate equation for
the time evolution of the site probabilities $(P_n(t))$ is the generalised master
equation (GME) [16,17]

$$\dot{P}_n = \int_0^t \sum_m [W_{mn}(t-t')P_m(t-t') - W_{nm}(t-t')P_n(t')]dt' \ . \tag{13}$$

In eq. (13) $W_{nm}(t-t')$ is the memory function responsible for transport between
sites n and m. The Markovian approximation to the memory function in eq. (13)
corresponds to a memory that is a delta function in time, $W_{nm}\delta(t-t')$. In this
limit eq. (13) reduces to the standard master equation for incoherent transport.
While the GME has been used extensively to bridge the gap between coherent and
incoherent motion [16,17], few applications of the GME to quantum transport in
disordered systems exist [18].

We now show that the nearest-neighbour memory function for our tight-binding
model can be used to study the Anderson transition on a Cayley tree of connectivity K. Using the procedure outlined by Zwanzig [16], it is straight-forward to
show that

$$W_{n,n\pm1}(t) = 2V^2\cos(\varepsilon_n-\varepsilon_{n\pm1})t \tag{14}$$

is the nearest-neighbour approximation to the memory function for our system. For
a 1-dimensional system then, the corresponding equation of motion for the site
probabilities is

$$\dot{P}_n = 2V^2\{\int_0^t \cos[(\varepsilon_n-\varepsilon_{n+1})(t-t')](P_{n+1}(t') - P_n(t'))dt'$$

$$+ \int_0^t \cos[(\varepsilon_n-\varepsilon_{n-1})(t-t')](P_{n-1}(t') - P_n(t'))dt'\} \tag{15}$$

Let us define the quantities $\Delta_m = (\varepsilon_{m+1} - \varepsilon_m)/\sqrt{2}V$ and $F_m(s) = s/(s^2+\Delta_m^2)$ where
$s = \varepsilon/\sqrt{2}V$, ε the Laplace variable conjugate to time. Noting that $F_m(s)$ is the
Laplace transform of eq. (14), we rewrite the equations of motion as

$$sP_n(s) - P_n(t=0) = F_n(s)(P_{n+1}(s) - P_n(s)) + F_{n-1}(s)(P_{n-1}(s) - P_n(s)) \tag{16}$$

We point out that in the Laplace domain eq. (16) resembles closely the Laplace-
transformed bond master equation (see eq. (13)) of the RH model. The only
difference is that the hopping rates $F_n(s)$ are frequency dependent, whereas the
ω_n's are static. Hence, the methods of section 3 can be used to develop and exact

t-matrix expansion [4] for the diffusion constant in the nearest-neighbour GME model. Recently, we explored such an expansion at the effective medium level in d-dimensions but were unsuccessful in predicting the Anderson transition in three dimensions [5].

A more promising approach that we have taken is an investigation of the energy spectrum of the Green function for the fluxes. In the frequency domain the singularities of the Green function determine the energy spectrum. If all the singularities are real simple poles, then the Green function describes a stationary system in which no transport obtains. Singularities with nonzero imaginary parts indicate the presence of branch points. This signifies the existence of delocalised states. Though the calculation of the energy spectrum is typically performed with the Green function for the amplitudes, the Green function for the fluxes and site probabilities, as we will see, suffice just as well to establish the existence of branch points.

To proceed, let us define the bond flux

$$J_{\underset{\sim}{n}}^{\alpha}(s) = F_{\underset{\sim}{n}-\hat{\alpha}}^{\alpha}(s)(P_n(s) - P_{n-\hat{\alpha}}(s)) \tag{17}$$

between sites n and $n-\hat{\alpha}$ along the α-direction. In the Laplace domain, the equations of motion in d-dimensions for the probabilities and fluxes are

$$sP_{\underset{\sim}{n}}(s) - P_{\underset{\sim}{n}}(t=0) = \sum_{\alpha=1}^{d} J_{\underset{\sim}{n}+\hat{\alpha}}^{\alpha}(s) - J_{\underset{\sim}{n}}^{\alpha}(s) \tag{18}$$

and

$$\frac{sJ_{\underset{\sim}{n}}^{\alpha}(s)}{F_{\underset{\sim}{n}}^{\alpha}} - \frac{J_{\underset{\sim}{n}}^{\alpha}(t=0)}{F_{\underset{\sim}{n}}^{\alpha}} = \sum_{\alpha=1}^{d} 2J_{\underset{\sim}{n}+\hat{\alpha}}^{\alpha}(s) + J_{\underset{\sim}{n}-\hat{\alpha}}^{\alpha}(s) - 2J_{\underset{\sim}{n}}^{\alpha} \tag{19}$$

respectively. The long-time properties of our system are determined along the axis $\mathrm{Re}(s) = 0$ or the $\mathrm{Im}(s)$ axis. If all the states are localised then the probability of remaining at the initial site (the origin) is either unity for all times or recurs to unity with a non-infinite recurrence time. For this state of affairs to obtain, all the singularities of the probability Green function must correspond to simple poles which lie along the $\mathrm{Im}(s)$ axis. From eqs. (18) and (19) it follows that the probability Green function G_P and the Green function for the flux at the origin G_J are related through

$$G_P^{\alpha}(\underset{\sim}{0},\underset{\sim}{0};s) = \frac{1}{s} - \frac{1}{s} \{G_J^{\alpha}(\underset{\sim}{0} + \hat{\alpha}, \underset{\sim}{0} + \hat{\alpha}; s) + G_J^{\alpha}(\underset{\sim}{0} - \hat{\alpha}, \underset{\sim}{0} - \hat{\alpha}; s)\} \tag{20}$$

where

$$G_J^{\alpha}(\underset{\sim}{n} + \hat{\alpha}, n + \hat{\alpha}; s) = [s^2 + \Delta_{\underset{\sim}{n},\alpha}^2 + 2 - \Gamma^{\alpha}(\underset{\sim}{n} + \hat{\alpha}; s)]^{-1} \tag{21}$$

is the return Green function for the flux between sites $\underset{\sim}{n}$ and $\underset{\sim}{n}+\hat{\alpha}$. In (21) $\Gamma^{\alpha}(n+\hat{\alpha};s)$ is the self energy along the α-direction for the flux between sites $\underset{\sim}{n}$ and $\underset{\sim}{n}+\hat{\alpha}$. It is evident from (21) and (20) that the singularities of the probability Green function $G_P^{\alpha}(\underset{\sim}{0},\underset{\sim}{0};s)$ cannot be simple poles along the $\mathrm{Im}(s)$ axis unless $\mathrm{Im}\Gamma^{\alpha}(\underset{\sim}{0}+\hat{\alpha};s)$ and $\mathrm{Im}\Gamma^{\alpha}(\underset{\sim}{0}-\hat{\alpha};s)$ are both identically zero. If we require that the probability distribution of $\mathrm{Im}\Gamma^{\alpha}(\underset{\sim}{0} + \hat{\alpha};s)$ and $\mathrm{Im}\Gamma^{\alpha}(\underset{\sim}{0} - \hat{\alpha};s)$ be equal, then $\mathrm{Im}\Gamma^{\alpha}(\underset{\sim}{0} + \hat{\alpha};s) = 0$ necessarily implies that $\mathrm{Im}\Gamma^{\alpha}(\underset{\sim}{0} - \hat{\alpha};s) = 0$ as well. Hence, we need only focus on the probability distribution of either $G_J^{\alpha}(\underset{\sim}{0}+\hat{\alpha}, \underset{\sim}{0}+\hat{\alpha};s)$ or $G_J^{\alpha}(\underset{\sim}{0}-\hat{\alpha},\underset{\sim}{0}-\hat{\alpha};s)$ to analyze the Anderson transition.

At this point we specialize to a Cayley tree to facilitate an exact calculation of the Green function

$$G_J(0,1,0,1;s) = [s^2 + \Delta_0^2 + 2 - \Gamma(0;s)]^{-1} \tag{22}$$

for the flux in the bond at the origin. We have dropped the superscript on $G_J(0,1,0,1;s)$ and $\Gamma(0;s)$ because directionality is not defined on a Cayley tree. By way of illustration, consider first the K=1 lattice. We number the sites $(\ldots-1,0,1,\ldots)$. The self-energy $\Gamma(0;s)$ is computed by summing over all closed skeleton paths that return to the origin. Clearly there are only two such paths: $0 \to 1 \to 0$ and $0 \to -1 \to 0$. The resultant self-energy at the origin

$$\Gamma(0;s) = \Gamma(0,0(1)) + \Gamma(0,0(-1)) \tag{23a}$$

$$= [s^2 + 2 + \Delta_{-1}^2 - \Gamma(-1,-1(0))]^{-1}$$

$$+ [s^2 + 2 + \Delta_1^2 - \Gamma(1,1(0))]^{-1} \tag{23b}$$

can be written as the sum of two self energies $\Gamma(0,0(1))$ and $\Gamma(0,0(-1))$. These quantities exclude all paths that visit sites 1 and -1, respectively. In general $\Gamma(n,n(m,\ell, \ldots))$ is the self-energy at site n which excludes all paths that visit sites $(m,\ell, \ldots)$. Self-consistency is introduced by requiring that the probability distribution for $\Gamma(0,0(1))$, $\Gamma(0,0(-1))$, $\Gamma(-1,-1(0))$ and $\Gamma(1,1(0))$ be equal. This self-consistency condition is justified because the probability distribution for a self energy of the form $\Gamma(n,n(m))$ should be independent of n and m. Now the signature of Anderson localisation is the vanishing of $Im\Gamma(0;s)$. Eq. (23a) implies that for $Im\Gamma(0;s) = 0$, $Im\Gamma(0,0(1)$ and $Im\Gamma(0,0(-1))$ must both vanish. However, because $\Gamma(0,0(1))$ and $\Gamma(0,0(-1))$ obey the same probability distribution, $Im\Gamma(0,0(1)) = 0$ necessarily implies that $Im\Gamma(0,0(-1)) = 0$ as well. Hence, the question of localisation can be answered in 1-dimension by investigating the properties of either $\Gamma(0,0(1))$ or $\Gamma(0,0(-1))$.

Consider then the defining equation

$$\Gamma(0,0(-1)) = [s^2 + 2 + \Delta_1^2 - \Gamma(1,1(0)]^{-1} \tag{24}$$

for $\Gamma(0,0(-1))$. We employed the Monte Carlo procedure of Abou-Chacra, Anderson and Thouless (AAT) [19] to solve Eq. (24) for the probability distribution of $Im\Gamma(0,0(-1))$. A flat distribution of 1000 members was used to generate random values for the site energies and $\Gamma(1,1(0))$. The mean value of $Im\Gamma(0,0(-1))$ was computed with s=iu and u chosen so that the mean of $Im\Gamma(0,0(-1))$ was maximised. The choice of u (s=iu) such that $Im\Gamma(0,0(-1))$ is maximized guarantees that when $Im\Gamma(0,0(-1))$ vanishes, all the fluxes are localized. We found that $Im\Gamma(0,0(-1))$ vanished for all values of W/V and for all values of u (even for infinitesimally small values of W/V). Hence, the nearest-neighbor memory function corroborates the well-known result in d=1 that all states are localised in the presence of disorder [20].

For K=2, the equation of motion for the flux at the origin involves the next-nearest neighbour fluxes as well. This state of affairs obtains because fluxes correspond to free bonds on a Cayley tree, each connected to 2K nearest-neighbour bonds. If one transforms the K=2 Cayley tree so that the bonds are represented as points, then the resulting structure is the expanded cactus shown in Fig. 1. On this lattice, the flux equations for a given site can be generated by summing over the vertices of the two triangles emanating from the flux site of interest. Here again the quantity of interest is the self-energy

$$\Gamma(0,0;s) = \Gamma(0,0(1,2)) + \Gamma(0,0(-1,-2)) \tag{25}$$

at the origin. As in the 1-dimensional case, we need only focus on one of the

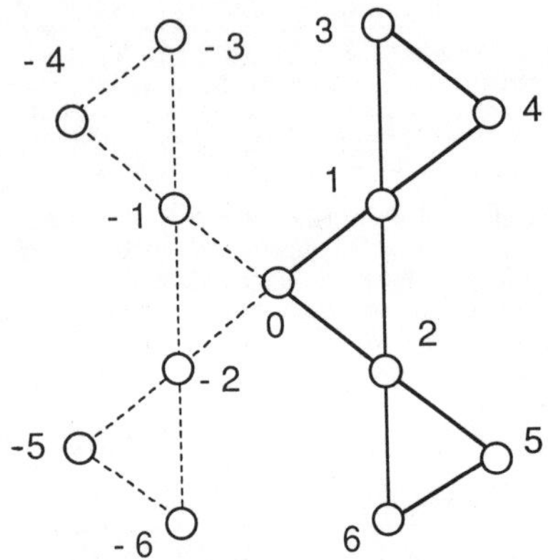

Figure 1. Expanded cactus for K=2. The fluxes are now lattice sites. In order to calculate the self-energy $\Gamma(0,0(-1,-2))$ on the expanded cactus, one must sum over all of the self-avoiding paths which exclude the "flux-sites" -1 and -2. There are three such paths $0 \to 1 \to 0$, $0 \to 2 \to 0$, and $0 \to 1 \to 2 \to 0$. These paths are decorated with paths over sites 3, 4, 5, and 6, etc. The entire left side of the figure is avoided for this calculation, as represented bvy the dashed bonds. Similarly, when calculating $\Gamma(0,0(1,2))$, the entire right side of the figure is avoided, i.e., none of the sites labeled by the positive integers are visited.

terms in (25). Consider $\Gamma(0,0(-1,-2))$. Three paths, $0 \to 1 \to 0$, $0 \to 2 \to 0$, and $0 \to 2 \to 1 \to 0$, contribute to the self-energy

$$\Gamma(0,0(-1,-2)) = [s^2 + 2 + \Delta_1^2 - \Gamma(1,1(0))]^{-1}$$

$$+ [s^2 + 2 + \Delta_2^2 - \Gamma(2,2(0))]^{-1}$$

$$- [s^2 + 2 + \Delta_2^2 - \Gamma(2,2(0))]^{-1}[s^2 + 2 + \Delta_1^2 - \Gamma(1,1(2,0))]^{-1} . \tag{26}$$

The self-energies that exclude one site can be expressed in terms of the self-energies that exclude two sites. The final self-consistent equation is

$$\Gamma(0,0(-1,-2)) = \frac{(s^2 + 2 + \Delta_2^2 - \Gamma(2,2(1,0))) + (s^2 + 2 + \Delta_1^2 - \Gamma(1,1(2,0))) - 1}{(s^2 + 2 + \Delta_2^2 - \Gamma(2,2(1,0))) \times (s^2 + 2 + \Delta_1^2 - \Gamma(1,1(2,0))) - 1} \tag{27}$$

Let us define $\Gamma(n,n(m,\ell)) = \Gamma_{nR} + i\Gamma_{nI}$ and $X_n = s^2 + 2 + \Delta_n^2 - \Gamma_{nR}$ for any site n and pair of sites (m,ℓ). The real and imaginary parts of $\Gamma(0,0(-1,-2))$ can be expressed in terms of these quantities. For $\text{Im}\Gamma(0,0(-1,-2))$ we obtain

$$\text{Im}\Gamma(0,0(-1,-2)) = \Gamma_{0I} = \frac{\Gamma_{2I}(\Gamma_{1I}\Gamma_{2I} + X_1^2 - X_1 + 1) + \Gamma_{1I}(\Gamma_{1I}\Gamma_{2I} + X_2^2 - X_2 + 1)}{(X_1 X_2 - \Gamma_{2I}\Gamma_{1I} - 1)^2 + (\Gamma_{2I}X_1 + \Gamma_{1I}X_2)^2} . \tag{28}$$

The imaginary s(s=iu) axis was scanned, and as is evident from Fig. 2 the maximum value of W/V needed to cause a transition was W/V = 13 at a value of u = 1.8. This result is in good agreement with the exact value of AAT of W/V = 17 [19]. The discrepancy between the value of AAT and the GME results stems of course from the $O(V^2)$ or nearest-neighbour approximation we have made to the memory function. This approximation effectively decreases the magnitude of the nearest-neighbor matrix element, V. Consequently, the electronic states become localized at a smaller

44

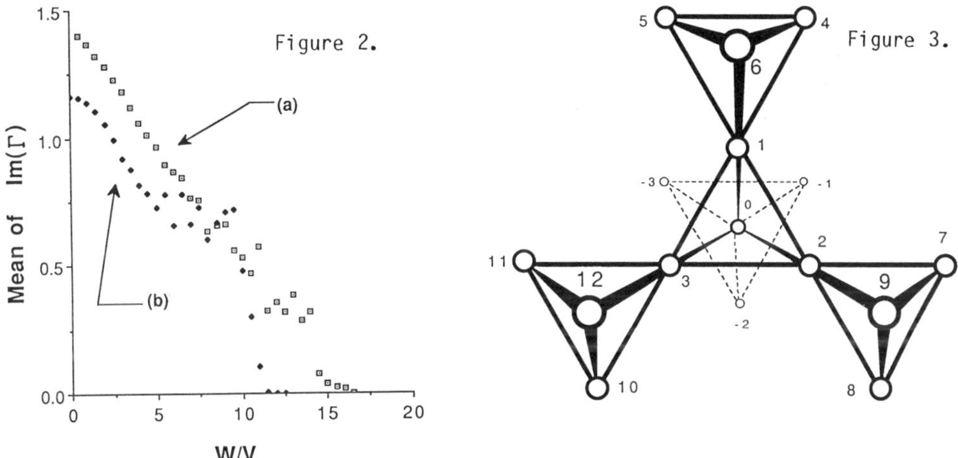

Figure 2. The geometric mean of the imaginary part of the self-energy $\Gamma(0,0)$, as a function of the ratio of the full-width of the disorder W and the nearest-neighbor overlap integral V, for the Cayley tree of connectivity K = 2. The points shown represent the average of the last 20 of 100 iterations of the self-consistent equation, with an ensemble of 500 members. The curves are (a) the self-energy for the amplitudes, which is generated from the exact expressions of AAT (8) with u = 0, and (b) the self-energy for the fluxes, which is generated from (4.14), with u = 1.8V. Curve (a) approaches zero at W/V = 17, while curve (b) approaches zero at W/V = 13.

Figure 3. Expanded cactus for the fluxes for K = 3 is a network of tetrahedrons, such that each "flux-site" is connected to six nearest-neighbors. In order to calculate $\Gamma(0,0(-1,-2,-3))$, it is necessary to sum over nine self-avoiding paths which completely exclude the sites labeled by the negative integers. The excluded sites are shown connected by dashed bonds. Three of the self-avoiding paths are the single-edge routes $0 \to 1 \to 0$, $0 \to 2 \to 0$, and $0 \to 3 \to 0$. Three others define the faces of the tetrahedron: $0 \to 1 \to 2 \to 0$, $0 \to 1 \to 3 \to 0$, and $0 \to 3 \to 2 \to 0$. The three remaining paths are, $0 \to 1 \to 2 \to 3 \to 0$, $0 \to 2 \to 3 \to 1 \to 0$, and $0 \to 3 \to 1 \to 2 \to 0$.

value of the disorder W. As we will see the K = 3 results also follow this pattern. It is likely then that inclusion of higher order terms in the memory function should improve the estimate of the critical value of W/V.

We now turn to the K=3 calculation. For K = 3, the equations of motion for the flux at site n can be obtained by summing over all the six nearest-neighbor fluxes of site n on the expanded cactus shown in Fig. 3. The lattice in Fig. 3 can be obtained from a K = 3 Cayley tree by decorating each lattice site with a tetrahedron. In analogy with the K = 2 calculation, the localisation-delocalisation transition can be studied from the self energy $\Gamma(0,0(-1,-2,-3))$. A closed form expression can be written down for $\Gamma(0,0(-1,-2,-3))$ by summing over the self-avoiding paths on the lattice shown in Fig. 2. We solved the resultant self-consistent equation for $\mathrm{Im}[\Gamma(0,0(-1,-2,-3))]$ using the Monte Carlo method of AAT. Our results are shown in Fig. 4. They indicate that the maximum value of $\mathrm{Im}\Gamma(0,0(-1,-2,-3))$ which occurs at u = 2.0 vanishes for W/V $\simeq$ 20 as opposed to the exact [19] value W/V $\simeq$ 29. Here again we see that the agreement between the GME with a nearest-neighbour memory kernel and the exact amplitude equation is excellent.

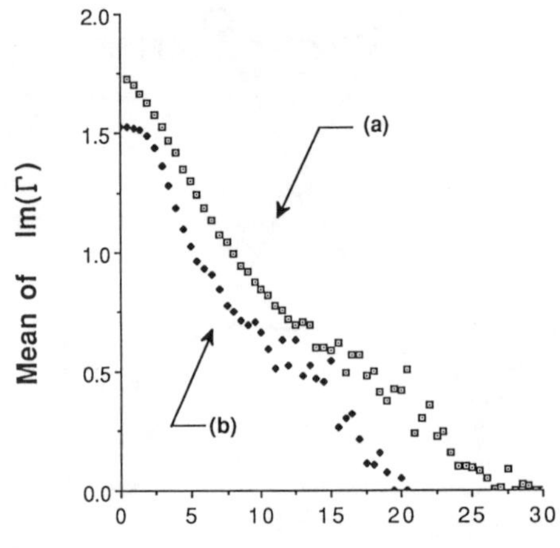

Figure 4.   The geometric mean of the imaginary part of the self-energy $\Gamma(0,0)$, as a function of the ratio of the full-width of the disorder W and the nearest-neighbor overlap integral V, for the Cayley tree of connectivity K = 3.   The points shown represent the average of the last 20 of 100 iterations of the self-consistent equation, with an ensemble of 500 members.   The curves are (a) the self-energy for the amplitudes, which is generated from the exact expressions of AAT (8) with u = 0, and (b) the self-energy for the fluxes, which is generated from (4.16), with u = 2.0V.   Curve (a) approaches zero quickly at W/V = 29, while curve (b) approaches zero at W/V = 20.

1.   K. Kundu and P. Phillips: Phys. Rev. A34, 707 (1987)
2.   K. Kundu, P. E. Parris and P. Phillips: Phys. Rev. B35, 3468 (1987)
3.   P. E. Parris, P. Phillips and K. Kundu: Physica A, in press
4.   P. E. Parris: Phys. Rev. B36, 5437 (1987)
5.   D. H. Dunlap, K. Kundu and P. Phillips: Phys. Rev. B, in press
6.   S. Alexander: Phys. Rev. B23, 2951 (1981); S. Alexander, J. Bernasconi, W. R. Schneider and R. Orbach: Rev. Mod. Phys. 53, 175 (1981)
7.   K. Kundu, D. Izzo and P. Phillips: J. Chem. Phys. 88, 2692 (1988)
8.   J. W. Haus, K. Kehr and J. W. Lyklema: Phys. Rev. B25, 2905 (1982)
9.   J. Machta, M. H. Ernst, H. van Beijeren and J. R. Dorfman: J. Stat Phys. 35, 413 (1984)
10. R. Zwanzig: J. Stat Phys. 28, 127 (1982)
11. N. G. van Kampen: Stochastic Processes in Physics and Chemistry (North-Holland, Amsterdam, 1981) p. 146
12. T. Odagaki and M. Lax: Phys. Rev. Lett. 45, 847 (1980); Phys. Rev. B25, 2301 (1982); 2307 (1982)
13. T. Odagaki: Phys. Rev. B, in press
14. R. Biller: Phys. Rev. B29, 2277 (1984). Biller uses a similar probability current approach for d=1.
15. P. W. Anderson: Phys. Rev. 109, 1492 (1958)
16. R. Zwanzig: In Lectures in Theoretical Physics, edited by W. E. Downs and J. Downs (Interscience, New York, 1961), Vol. 3
17. V. M. Kenkre: In Exciton Dynamics in Molecular Crystals and Aggregates, edited by K. G. Hohler (Springer, Berlin, 1982)
18. C. Aslangul, N. Pottier and D. Saint-James: preprint.
19. R. Abou-Chacra, P. W. Anderson and D. J. Thouless: J. Phys. C: Solid St. Phys. 6, 1734 (1973)
20. E. Abrahams, P. W. Anderson, D. C. Licciardello and T. V. Ramakrishnan: Phys. Rev. Lett. 42, 673 (1979)

Nonlinearity and Randomness in Quantum Transport

V.M. Kenkre

Department of Physics and Astronomy, University of New Mexico,
Albuquerque, NM 87131, USA

Strong interactions with lattice vibrations can lead to profound modifications in the transport of
quasiparticles such as electrons or excitons moving in solids. Recent approaches to the problem of the
description of such transport have been based on discrete nonlinear quantum evolution equations. This
is a brief review of some aspects of this recent activity. The first part of the article consists of a bird's-
eye-view presentation of several recent results for the nonlinear quantum dimer as well as for spatially
extended systems, and of their applications to various experiments ranging from muon spin relaxation
to fluorescence depolarization. The second part focuses attention on a particular kind of the interplay
of nonlinearity and of randomness in such systems.

1. Introduction

The problem of the description of strong interactions of lattice vibrations with quasiparticles moving in
a lattice has had a long history and has been approached via a variety of different methods [1-15]. The
most recent methods of attack [5-15] consist of analyzing the quasiparticle transport via nonlinear
equations, the source of the nonlinearity being the strong interactions with the lattice. Several different
sets of such transport equations have been arrived at from microscopic analysis or postulated[5-10].
Here we consider the simplest version appropriate to our purposes:

$$i\, dc_m/dt = V(c_{m+1} + c_{m-1}) - \chi\, |c_m|^2 c_m \tag{1.1}$$

which describes the evolution of the amplitude $c_m(t)$ for occupation of the site m by the
quasiparticle, the (nearest-neighbour) intersite transfer matrix element being V and the nonlinearity
parameter, which measures the lowering of the site energy, being χ. The details of how one arrives at
(1.1) from a microscopic starting point are not the subject of the present discussion. A simple way
consists of applying standard time scale disparity arguments to the coupled equations introduced by
Scott and collaborators[6-8]. Generalizations of (1.1) such as those to higher-dimensional lattices, or
to longer-range transfer interactions, are straightforward.

The focus of the work reviewed below is the investigation of the consequences of (1.1) in
various physical situations, and the application of the results to specific experiments. Results we have
chosen for a presentation here may be divided into three categories: numerical results for the extended
chain, approximate analytical results for the extended chain, and exact analytical results for the dimer.
The first two are described in section 2 and the third in section 3 below. Analytical development aimed
at the description of the interplay of randomness and nonlinearity forms the content of section 4, and
conclusions are presented in section 5.

2. Results for Extended Chains

Exact analytic solutions of (1.1) are not available for the extended chain, its closest soluble relative being the Ablowitz equation to be found in refs. [16]. While work is under way on the application of the latter to observations, our concern here is (1.1). We present two kinds of results for the extended chain: numerical results for (1.1) as applied to muon spin relaxation, and a perturbative analytical procedure which treats the nonlinearity exactly but the site-to-site transfer approximately.

2a: Numerical Results and Application to Muon Spin Relaxation

A modern probe in solid state physics consists of spin relaxation observations of muons introduced into a solid for the purpose of the measurement of various properties of the solid. In ferromagnetic crystals with magnetically inequivalent sites, e.g. BCC iron, a spin-polarized muon undergoes spin relaxation as a combined result of its motion and of the different Larmor frequencies at the different sites. An idealization of this system is a linear chain with alternating Larmor frequencies ω and $-\omega$. In such a simplified system, the muon may be looked upon as a quantum particle whose state evolves under the combined action of transport via nearest-neighbour matrix elements V from site to site, spin rotation under the Larmor frequencies, and strong interaction with the vibrations of the lattice. Kehr and Kitahara[17] studied this system and assumed that the last mentioned effect, viz. the dynamical disorder caused by the strong interactions with the lattice, could be described by the so-called stochastic Liouville equation[4]. A natural question to ask is what effects are likely to be observed in muon spin relaxation if muon transport obeys the nonlinear discrete Schrödinger equation (1.1) which includes, in its description, (nonlinear) feedback effects, rather than the stochastic Liouville equation, which is not capable of describing such effects.

As in the stochastic Liouville equation analysis[17], we calculate[18] the x-component of the muon spin given by

$$S_x(t) = \Sigma_m \; \tfrac{1}{2} \, [\, \rho_{mm}{}^{+\,-}(t) \; + \; \rho_{mm}{}^{-\,+}(t)\,] \tag{2.1}$$

The summation is over all the sites of the lattice (linear chain in the present case) and the quantities summed are the density matrix elements between the + and - states at the same site. The + and - states are the Zeeman-split states differing in energy by Δ (which is proportional to an applied static magnetic field). The muon is initially localized at a single site ($m=0$) and occupies the + and - states with equal probability and phase. For a full description of the equations of motion we refer the reader to refs [17,18]. As Kehr and Kitahara have shown[17], the linear case, i.e. the one in which (1.1) applies with $\chi = 0$, gives, for $S_x(\epsilon)$, the Laplace transform of $S_x(t)$:

$$S_x(\epsilon) = (1/2\epsilon) \, \{ \, 1 - \Delta^2 \, [\, (\Delta^2 + \epsilon^2)(\Delta^2 + 16V^2 + \epsilon^2) \,]^{-\frac{1}{2}} \tag{2.2}$$

which, when inverted, shows mild oscillations of S_x from its initial value of 0.5 and eventual decay to a value lower by the factor $\{ \, 1 - [\, 1 + (4V/\Delta)^2 \,) \,]^{-\frac{1}{2}}$. In fig. 1a we show this linear behaviour contrasted with the nonlinear predictions of (1.1) which we obtain numerically. Wild oscillations of the muon spin component are seen along with distinct structure in the evolution. The equilibrium value is reduced as a result of the nonlinearity. These features differ appreciably from the damping behaviour seen in the stochastic Liouville equation analysis[17]. The physical origin of our results is understood easily: the structure in the evolution arises from the apparent non-degeneracy and additional energy

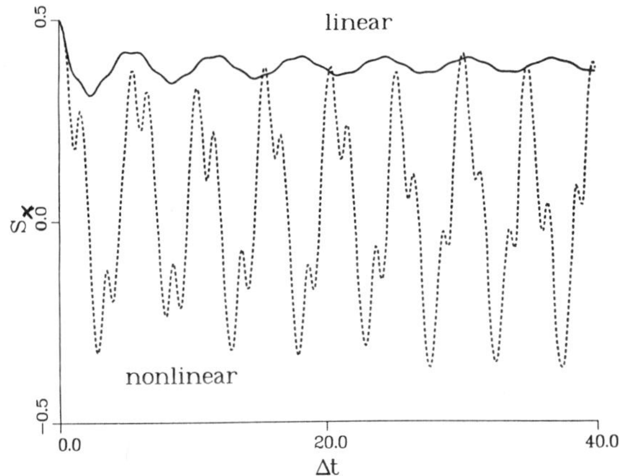

Fig. 1a Effect of transport nonlinearity on muon spin relaxation: shown is the time-dependence of the x-component of the muon spin, in the linear and the nonlinear cases. One sees the reduction in the muon mobility and the appearance of structure arising from the energy mismatch

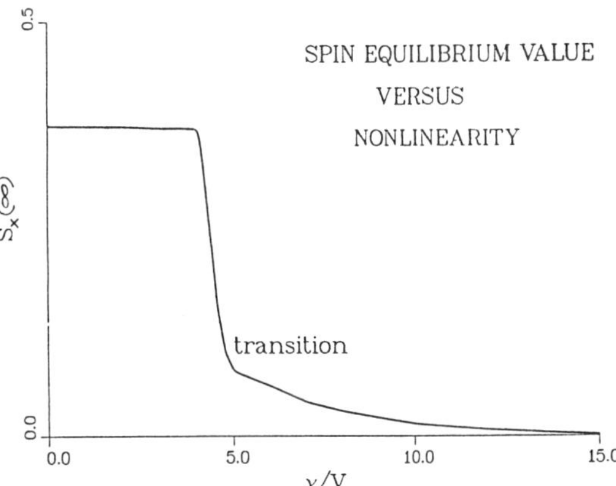

Fig. 1b Abrupt transition in the equilibrium value of the x-component of the muon spin as a function of the nonlinearity

mismatch imposed by the nonlinearity, while the reduction of the equilibrium value is representative of the reduction of the overall mobility of the quantum particle brought about by the strong lattice distortion which accompanies it during its motion.

Our predictions from the nonlinear evolution equation can be appreciated further from fig. 1b in which we plot the equilibrium value (as $t \to \infty$) of S_x as a function of the nonlinearity ratio χ/V which measures the strength of the interaction with the lattice relative to the intersite transfer. An abrupt transition is evident as the nature of the particle motion changes from being predominantly free to predominantly self-trapped.

2b: Approximation Procedure Based on a Perturbative Approach

Returning to analytical procedures, it is straightforward to cast (1.1) into a nonlinear Liouville-Von Neumann equation for the density matrix ρ of the quasiparticle:

$$d\rho/dt = -i L_V \rho + i L_\chi \rho = -i L \rho \tag{2.3}$$

where the "Liouville" operators L_V and L_χ are defined through

$$L_V O = [V, O] \tag{2.4}$$

$$(L_\chi O)_{mn} = \chi\,(\rho_{mm} - \rho_{nn})\,O_{mn} \tag{2.5}$$

for any operator O. The application of diagonalizing projection operators[19] P and $(1-P)$ to (2.3) yields coupled equations for the diagonal part ρ' and the off-diagonal part ρ'' of the density matrix. Although (2.3) is nonlinear, the specific nature of the definition (2.5) makes the equation for ρ'' linear and thereby allows the use of standard methods towards the solution of ρ'' as in the usual linear case. The result[20] for initial random phases is the generalized master equation (GME) obeyed by the probabilities $P_m(t)$:

$$dP_m(t)/dt = \int_0^t ds\ \Sigma_n\ [\ \mathcal{W}_{mn}(t,s)P_n(s) - \mathcal{W}_{nm}(t,s)P_m(s)\ ] \tag{2.6}$$

where the nonlinear memory functions $\mathcal{W}_{mn}((t,s)$ are obtained from the elements of the tetradic

$$PL_V \int ds\ \{\exp[-i\int dz(1-P)L(z)]\}(1-P)L_V(s)\rho'(s).$$

Since our interest is primarily in the nonlinearity, we do not treat it perturbatively. On the other hand, the site-to-site transfer may be considered to be weak enough to be analyzed via an approximation. We thus aply to (2.8) the standard weak-coupling appoximation wherein L_V is taken to have a small effect compared to the effect of the nonlinearity. An expansion of the exponential in powers of $-i\int dz\ L_\chi(z)$ and the use of the definition of L_χ gives

$$(\{\exp[-i\int dz\ L_\chi(z)]\}\ O)_{mn} = (\exp(-i\int dz\chi[\ \rho_{mm}(z)-\rho_{nn}(z)]\})\ O_{mn} \tag{2.7}$$

It is then straightforward to evaluate the nonlinear memory functions as

$$\mathcal{W}_{mn}(t,s) = 2V^2\,(\delta_{m,n+1} + \delta_{m,n-1})\,\cos\{\ \chi \int_s^t dz\ [\ P_m(z)-P_n(z)]\} \tag{2.8}$$

Equation (2.6) along with (2.8) is the nonlinear generalized master equation in the weak-coupling approximation for the intersite transfer. The approximation inherent in (4.2) is of a weak-coupling nature in V but not in the nonlinearity χ. The latter is taken to infinite order and the former to the lowest non-vanishing order. As we shall point out in section 4 below, the memory functions deduced above are exact for the case of a two-site system. For extended systems the weak-coupling GME (2.8) inherits the features of its linear counterpart. It is well known[2,4] that the weak-coupling memory functions in the linear case produce negative probabilities but yield the exact mean-square displacement or velocity auto-correlation function. Thus, while calculations of probabilities or related quantities with the weak-coupling approximation must be undertaken with care, the above treatment is well suited to the calculation of moments of probability, mean-square displacements, velocity correlation functions, diffusion constants, and transfer rates.

Our numerical calculations of such quantities as the mean-square displacement and suitably defined effective transfer rates[20], have shown that the qualitative behaviour of the time dependence, *including the transition,* is retained by our approximate GME ,and that the difference is large for values of the nonlinearity χ near the transition from free to self-trapped motion.

3. <u>Exact Results for the Nonlinear Quantum Dimer</u>

While it is true that exact solutions (1.1) for a chain of arbitrary number of sites are not known, the complete analytic solution has been obtained for a dimer, i.e. a two-site system. Much work[7,10-15] has been recently done on the basis of that solution. It includes the exact time evolution in terms of Jacobian elliptic functions for arbitrary initial conditions, [10,11], explicit demonstration of self-trapping transitions and other rich behaviour in the time evolution[10-13], evaluation of the stationary self-trapped states of the dimer [7,11,12], application of the analysis to specific experimentally realizable dimer systems[11,14] and evaluation of nonlinear memory functions[20,21].

3a. The Complete Analytic Solution

The form of (1.1) for a dimer is

$$dc_1/dt = V\, c_2 - \chi\, |c_1|^2\,; \qquad dc_2/dt = V\, c_1 - \chi\, |c_2|^2 \qquad (3.1)$$

The complete analytic solution of (3.1) can be obtained[10,11] by deriving the corresponding density matrix equations, and extracting and solving a closed equation for the probability difference between the two sites. If we define the quantities p, r, and q through

$$p = \rho_{11} - \rho_{22}\,; \qquad r = \rho_{12} + \rho_{21}\,; \qquad q = \rho_{12} - \rho_{21} \qquad (3.2)$$

we find that r and q are determined from p through

$$r(t)=r(0)+(\chi/4V)[p^2(t)-p^2(0)]\,; \qquad q(t)=q(0)+(1/2iV)[p(t)-p(0)] \qquad (3.3)$$

and that the probability difference p obeys the closed nonlinear equation

$$d^2p/dt^2 = A\, p - B\, p^3 \qquad (3.4)$$
$$A = (\chi^2/2)p^2(0) - 4V^2 - 2V\chi\, r(0)\,; \qquad B = (\chi^2/2) \qquad (3.5)$$

The full solution of the nonlinear equation (3.4) is[11]

$$p(t) = C\ \mathrm{cn}\ [(C\chi/2k)(t-t_0)\,|\,k] = C\ \mathrm{dn}\ [(C\chi/2)(t-t_0)\,|\,1/k] \qquad (3.6)$$

The arbitrary constant C and the elliptic parameter k satisfy

$$C^2 = p^2(0) - \zeta^2 + [\zeta^4 - (4V/\chi)^2\, q^2(0)]^{1/2} \qquad (3.7)$$

$$1/k^2 = 2\,[\,1 + (1/C^2)(\zeta^2 - p^2(0)\,)\,] \qquad (3.8)$$

where ζ is given by

$$\zeta^2 = (1/2)[\,(4V/\chi)^2 + (8V/\chi)r(0)\,] \qquad (3.9)$$

The starting time t_0 is usually zero but can be written in terms of the normal elliptic integral of the first kind in terms of C and k[11].

3b. Transitions and Self-Trapping Phenomena

Although mathematically simple, the treatment of the dimer based on the solution given above is rich in physical insights and elucidates at least three separate transitions as well as phenomena such as self-trapping. The first of the transitions is a "static transition" to be seen in the stationary states of the dimer. Observed first in ref. [7], the stationary states can be obtained from our analysis above by equating the right hand side of (3.4) to zero[11,12]. It is seen that while the stationary states involve equal occupation on the two sites if $\chi < 2V$, self-trapping, represented by the unequal occupation

$$|c_{1,2}| = (1/2)^{\frac{1}{2}} \{ 1 \pm [1 - (2V/\chi)^2]^{\frac{1}{2}} \}^{\frac{1}{2}} \tag{3.10}$$

occurs when χ exceeds the critical value 2V.

For the initially localized condition, one finds a different transition in the *time evolution* and an additional indication of self-trapping. The values of C and p(0) in (3.6) are now 1, $k = 4V/\chi$, and the solution for $p(t)$ is

$$p(t) = cn [2Vt \mid \chi/4V] \tag{3.11}$$

The time evolution shows a slowing down as the nonlinearity is increased, a "dynamic" transition at $\chi = 4V$ when $p(t)$ reduces to a sech curve which is non-oscillating , and a change to oscillations that do not cross the $p = 0$ region as χ exceeds the critical value $4V$. This last effect is yet another manifestation of self-trapping.

A third transition is observed[12,13] in the time evolution for non-localized initial conditions as a result of the interplay of phases and nonlinearity. If we consider, for simplicity, the case in which the initial amplitudes are real and have opposite signs, $r(0)$ and $p(0)$ are related through

$$r(0) = -[1-p^2(0)]^{\frac{1}{2}} \tag{3.12}$$

For χ values which are large enough (in excess of those required for the self-trapping "dynamic"
transition similar to the one explained above) it is now possible to satisfy $\chi/4V = (1/2) [1-p^2(0)]^{\frac{1}{2}}$.
For this choice of the degree of nonlinearity and initial conditions, k is infinite, the system finds itself initially in one of its stationary states, and the probability of occupation of the sites remains unchanged from its initial value. A further increase in χ makes k imaginary. An application of the imaginary argument transformation to the cn function to get the nc function, followed by that of the reciprocal modulus transformation to the nc function, shows that $p(t)$ is described by the Jacobian nd function. A striking characteristic of this transition is that, beyond the critical value of χ, a "repulsion" effect[12] is exhibited in which the probability on the site with initially larger (smaller) occupation *increases* (*decreases*) in occupation and remains larger (smaller) than its initial value.

3c. <u>Application to Experiments: Fluorescence Depolarization</u>

Applications of the dimer analysis have been made to several experimental situations. They include, among others, neutron scattering off hydrogen atoms trapped at impurity sites in metals[11], and the phenomenon of sensitized luminescence. We will examine here, briefly, a recently reported application[13] to fluorescence depolarization[22]. The system under investigation is a variable-distance noninteracting donor-acceptor pair of molecules. A practical example is provided by the so-called "stick-dimers" in which poly-L-proline oligomers of controllable length are used to separate an â-napththyl group at the carboxyl end - the donor - from the dansyl group at the imino end - the acceptor-, and the efficiency of energy transfer is studied through measurements of fluorescence excitation, emission and polarization spectra. On illumination, either of the molecules in the pair, assumed identical to each other in the simplest version of the analysis, may undergo electronic excitation. The direction of the induced dipole moment produced on the molecule through the process of excitation depends on geometrical factors and is generally different for the two molecules in the pair. For simplicity, we will assume the two dipole moments to be mutually perpendicular. One could, in principle, create an excitation of the dimer which is localized on one of the two molecules by shining (broad-band) light polarized in the direction of the dipole moment on that molecule. Varying the angle of the polarization of the incident light beam would result in varying the relative amplitude or probability of excitation of either molecule. If $I_\parallel$ and $I_\perp$ are the intensities of fluorescence polarized respectively parallel and perpendicular to the direction of the polarization of the incident light, the degree of fluorescence polarization f, given by

$$f = (I_\parallel - I_\perp) / (I_\parallel + I_\perp) \tag{3.13}$$

is a convenient experimental observable for the investigation of excitation transfer within the dimer. It can be shown that, with the understanding that ϕ is the angle made by the polarization of the incident light with the induced dipole moment on molecule 1, the degree of fluorescence depolarization f is given in terms of the density matrix elements of the dimer as

$$f = p \cos 2\phi + r \sin 2\phi \tag{3.14}$$

In fig. 2 we plot $f_s(\phi)$, the steady state value of f (weighted by the decay factor $\exp(-t/\tau)$ where τ is the lifetime), as a function of the polarization angle ϕ for various degrees of the nonlinearity parameter χ. For $\chi/4V = 0$, the system is linear and the integrated signal varies sinusoidally as a function of the polarization angle ϕ. As χ takes on non-zero values, minor changes occur in this variation for small nonlinearities. However, beyond a critical value of χ, the curve changes qualitatively and pronounced new effects are seen around $\phi = 135^0$. The changes brought about by nonlinearity are trivial until the ratio $\chi/4V$ equals $\frac{1}{2}$. Peculiar dips now appear in the f-curve and there seems to be a "pinning" of the curve at 135^0. Curves a,b,c,d, e represent various values of the nonlinearity ratio $\chi/4V$: 0, 0.2, 0.5, 1, and 1.5. Curve a $(\chi = 0)$ shows the simple sinusoidal variation characteristic of a linear dimer. Curve b shows a trivial change in the variation brought about by the nonlinearity while the dimer is still in the "free" region. Curve c $(\chi = 2V)$ shows the transition to self-trapped behavior. A flattening of the dependence of f on ϕ is characteristic of the transition and is apparent in curve c. Curves d and e show the dramatic effects of nonlinearity: the shift of the maximum away from 135^0 and the "pinning" at 135^0.

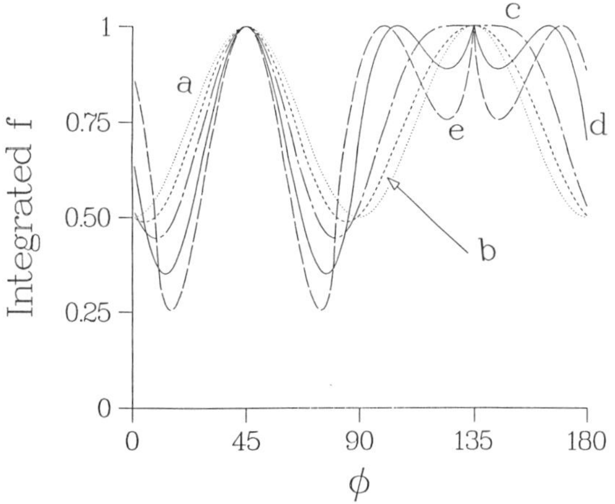

Fig. 2   Nonlinear effects on fluorescence depolarization: the steady state f plotted as a function of ϕ, the angle of polarization (in degrees) for various values of $\chi/4V$, the nonlinearity ratio. A shift of the maximum and a cusp at $\phi = 135$ degrees are the pronounced effects of non-linearity. See text for values of the nonlinearity.

The consequences of nonlinearity are, thus, (i) the slight variation of $f(\phi)$ with χ in the region $\chi < 2V$, (ii) the transition at $\chi = 2V$, (iii) the shift of the maximum away from $\phi = 135^0$, and (iv) the cusp at 135^0. For want of space we do not discuss the detailed origin of these effects here but refer the reader to ref. [13].

4. Interplay of Nonlinearity and Randomness

We now turn to a brief description of a certain interplay of randomness and nonlinearity[23]. Unlike standard treatments in this area, we do not take the static site energies or the intersite transfer matrix elements to be random. Instead, we investigate the effect of random nonlinearities, i.e. situations in which χ is itself random. The nonlinearity parameter depends for its value on the magnitude of the interaction of the moving particle with the vibrations of the lattice and on the frequencies of the vibrations. These frequencies and the coupling constants could vary from site to site and be therefore regarded as random parameters, the situation being analogous to inhomogeneous broadening of spectral linewidths or dephasing of spins in nuclear magnetic resonance experiments. Our system is an ensemble of dimers in which the χ's are given by a distribution $p(\chi)$. Our approach is to ensemble average the nonlinear memory functions obtained in the manner illustrated in section 2b above.

The analysis of section 2b, when applied to the dimer, results in the nonlinear GME (2.6) with symmetric memory functions given by[20]

$$\mathcal{W}_{12}(t,t') = \mathcal{W}_{21}(t,t') = \mathcal{W}(t,t') = 2V^2 \cos \left\{ \chi \int_{t'}^{t} ds \, p(s) \right\} \tag{4.1}$$

which are similar to those given in (2.8) in form but differ from them in that (4.1) does *not involve any* weak-coupling approximation. An ensemble-average of the *exact* memory functions over the distribution of χ's, i.e.,

$$\langle \mathcal{W}(t,t') \rangle = \int d\chi \, p(\chi) \, \mathcal{W}(t,t') \tag{4.2}$$

will now be used as the effective memory function in the GME. Equations (4.1) and (4.2) show that $\langle W(t,t') \rangle$ is essentially a Fourier (cosine) transform of the distribution function p (not to be confused with the density matrix), the variable transformed being χ, and the transform-variable ξ being related to the probability difference p through

$$d\xi(t)/dt = p(t) \tag{4.3}$$

The GME can now be written as an evolution equation not for p but for its integral ξ:

$$d^2\xi/dt^2 + \Omega(\xi) = 0 \tag{4.4}$$

which may be looked upon as the evolution equation of a fictitious oscillator for which the sum of its "potential energy" $U(\xi)$, which equals $\int d\xi \, \Omega(\xi)$, and of its "kinetic energy" which equals half the square of $\langle d\xi/dt \rangle$, is a constant of the motion.

A number of interesting manifestations of the interplay of randomness and nonlinearity may now be obtained[23] with the help of (4.4) or its "potential diagram". If the ensemble distribution $p(\chi)$ is a δ-function at χ_0, we recover the results of section 3, including the pendulum equation derived by Cruzeiro-Hansson[15], obtain for the "force" $\Omega(\xi)$, the expression

54

$$\Omega(\xi) = (4V^2/\chi_0)\sin(\xi\chi_0) \qquad\qquad (4.5)$$

and find that the potential $U(\xi)$ is proportional to a cosine. We plot this potential in fig. 3a and see that, for this "δ-ensemble", the system point escapes in the self-trapped case (ξ is unbounded for large times which means that the probability difference p does not oscillate around a vanishing value) but is bound in the free case (ξ is constrained to oscillate around the value zero). For a distribution $\rho(\xi)$ consisting of a square pulse of width Δ centered around χ_0, (Δ and χ_0 being respective measures of the degree of randomness in the nonlinearity, and of its average value), a new feature emerges. The "force" is given by

$$\Omega(\xi) = (4V^2/\chi_0)\{\ \text{si}[\xi(\chi_0 + \tfrac{1}{2}\Delta)] - \text{si}[\xi(\chi_0 - \tfrac{1}{2}\Delta)]\ \} \qquad\qquad (4.6)$$

where si is the sine integral function. In fig. 3b we plot the potential $U(\xi)$ obtained by integrating $\Omega(\xi)$ and see that, for this "pulse-ensemble", the amplitude of the potential is damped as ξ increases in magnitude. In the self-trapped case, this has the obvious result that, as the system point in ξ-space escapes (since p, whose integral ξ is, does not vanish or oscillate around zero), the "velocity" of the system point tends to a constant. This ξ-velocity is $d\xi/dt$, i.e. p itself. The effect of randomness in this "pulse-ensemble" is to cause the probability difference to be damped for large times in the self-

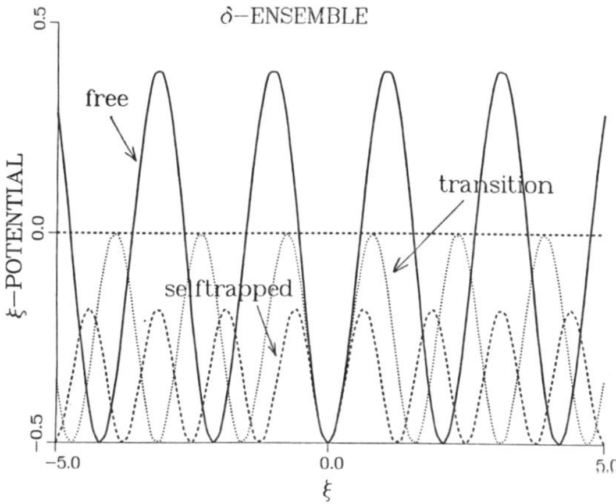

Fig. 3a   The potential $U(\xi)$ obtained by integrating $\Omega(\xi)$ from (4.5) plotted as a fun-ction of $\xi = \int dt\, p(t)$ for the case of no randomness (δ-ensemble). The free, the self-trapped, and the transition case are shown. Escape of the ξ-point represents self-trapping and its being constrained describes free motion.

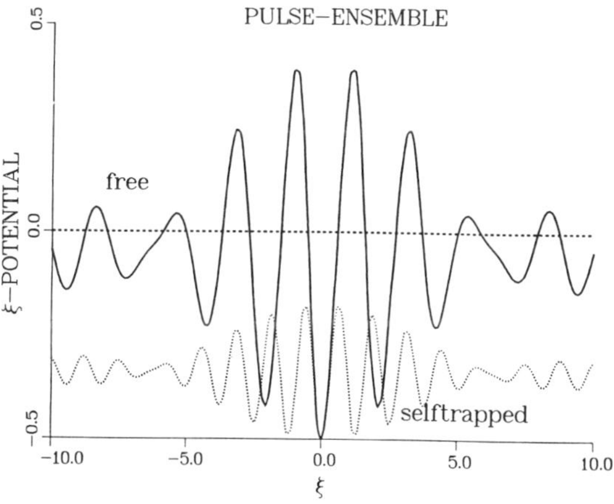

Fig. 3b The potential $U(\xi)$ obtained from (4.6) plotted for the case of a pulse-ens-emble, with finite values of both the average and width of the nonlinearity χ. Randomness has only a slight effect in the free case but causes damping of the self-trapped state. This is clear from the fact that $U(\xi)$ ceases to oscillate for large ξ.

trapped case for which χ_0/V, the average value of the nonlinearity, is larger than a critical value. No damping occurs below the critical value, i.e. when the motion is free.

A number of different ensembles have been studied in this manner[23]. The Lorentzian and Gaussian ensembles destroy self-trapping completely. The Poisson distribution $P(\xi) = (\chi^n/n!)\exp(-\alpha\chi)$ has the same effect for $n=1$ but not for $n>1$. The general conclusions to be drawn for this particular kind of interplay of disorder and nonlinearity, are that randomness generally opposes self-trapping, that in some cases it eliminates self-trapping, and that in others it causes damping of the self-trapped state.

We mention in passing that the generalization of the above analysis of the interplay of randomness and nonlinearity valid to a *chain* rather than a dimer may be undertaken by applying ensemble averages to the weak-coupling GME (2.8). For a Lorentzian distribution P, one obtains[20] the generalization of (4.4) as

$$d^2\xi_m/dt^2 = (\pi V^2/\alpha)\, [\,\exp(-\alpha|\xi_m - \xi_{m-1}|) - \exp(-\alpha|\xi_{m+1} - \xi_m|)] \qquad (4.7)$$

which is remarkably similar to the evolution equation for the Toda lattice[24]:

$$d^2\xi_m/dt^2 = (const)\{\,\exp[-\alpha(\xi_m - \xi_{m-1})] - \exp[-\alpha(\xi_{m+1} - \xi_m)]\} \qquad (4.8)$$

the only difference between (4.6) and (4.7) being that the absolute values of the differences of ξ_m at neighbouring sites appear in the former while their actual values appear in the latter.

5. <u>Concluding Remarks</u>

The essential feature of the systems of interest to the present investigation is a strong interaction between a quasiparticle which moves on a lattice in keeping with a quantum evolution equation, and oscillators whose displacements modulate the quasiparticle parameters such as the site energy. During the conference whose proceedings form the present volume, it was pointed out that almost all the conference lectures dealt with nonlinearity and disorder as separate entities rather than subjects in interplay. The present discussion shares that trait. An attempt was made, however, (in the lecture as well as the article) to focus on the interplay of nonlinearity and disorder in one part, viz. section 4.

The interaction of quasiparticles or excitations in solids with molecular or lattice vibrations can lead to a number of strong and discernible effects on transport phenomena. Traditionally, these interactions have been studied with the help of scattering, dressing and polaronic concepts[1-4]. A new theoretical tool has appeared in the literature recently: the nonlinear Schrödinger equation pioneered for excitation transport by Davydov and others[5,6]. We believe it is of crucial current importance to explore the consequences of the interaction of excitations with vibrations with the help of this tool, particularly in the context of experimental observations. This lecture/review has attempted to describe some of our work in these directions. There are several other aspects of the work, such as the subject of rotational polarons[25] (the exploration of consequences of *nonlinear* quasiparticle-lattice interactions and of *inelastic* restoring forces) and of non-adiabatic evolution(the investigation of the consequences of lattice relaxation which is *not infinitely fast*). These are being actively pursued at the moment but have not been reported on for want of space.

<u>Acknowledgements</u>

It is a pleasure to thank Mr. Honglu Wu, Dr. George P. Tsironis,Dr. David Campbell, and Prof. Mark Ablowitz for discussions, Prof. Klaus Kehr for making his work available before publication, and Mr. Salil Prabhav Niman for general assistance. This work was supported in part by the National Science Foundation and the Department of Energy under respective grants DMR-850638 and DE-FGD4-86ER45272.

<u>References</u>

1. T. D. Holstein, Ann Phys. (NY) <u>8</u>, 325, 343 (1959); D. Emin, Adv. Phys. <u>24</u>, 305 (1975)
2. R. J. Silbey, Ann. Rev. Phys. Chem. <u>27</u>, 203 (1976)
3. C. B. Duke and L. B. Schein, Physics Today <u>33</u>, no. 2, 42 (1980)
4. V. M. Kenkre and P. Reineker, <u>Exciton Dynamics in Molecular Crystals and Aggregates</u>, ed. G. Höhler (Springer-Verlag, Berlin, 1982)
5. A. S. Davydov, J. Theor. Biol. <u>38</u>, 559 (1973); Usp. Fiz. Nauk. <u>138</u>, 603 (1982) and references therein
6. A. C. Scott, F. Y. Chu, and D. W. McLaughlin, Proc. I.E.E.E.<u>61</u>, 1443 (1973)
7. J. C. Eilbeck, P. S. Lomdahl and A. C. Scott, Physica D <u>16</u>, 318 (1985)
8. J. M. Hyman, D. W. McLaughlin and A. C. Scott, Physica D <u>3</u>, 28 (1981)
9. D. W. Brown, K. Lindenberg and B. J. West, Phys. Rev. B <u>37</u>, 2946 (1988)
10. V. M. Kenkre and D. K. Campbell, Phys. Rev. B <u>34</u>, 4959 (1986)
11. V. M. Kenkre and G. P. Tsironis, Phys. Rev. B <u>35</u>, 1473 (1987)
12. V. M. Kenkre, G. P. Tsironis and D. K. Campbell, in <u>Nonlinearity in Condensed Matter</u>, ed. A. R. Bishop, D. K. Campbell, P. Kumar and S. E. Trullinger, (Springer-Verlag 1987)
13. G. P. Tsironis and V. M. Kenkre, Phys. Letters A <u>127</u>, 209 (1988)
14. V.M.Kenkre and G.P.Tsironis, Chem. Phys., to be published, Dec. (1988)
15. L. Cruzeiro-Hansson, P. L. Christiansen and J. N. Elgin, Phys. Rev. B <u>37</u>, 7896 (1988)
16. M. Ablowitz, SIAM, Review <u>19</u>, 663 (1977); M. J. Ablowitz and J. F. Ladik, J. Math. Phys. <u>17</u>, 1011 (1976)
17. K. W. Kehr and Kazuo Kitahara, J. Phys. Soc. Japan, <u>56</u>, 889, 4289 (1987); K. W. Kehr and Kazuo Kitahara, to be published.
18. V. M. Kenkre and H. -L. Wu, University of New Mexico preprint
19. R. Zwanzig, in <u>Lectures in Theoretical Physics</u> vol.III, ed. W. Brittin, B. Downs and J. Down (Interscience 1961)
20. H. -L. Wu and V. M. Kenkre, Phys. Rev. B <u>39</u>, xxxx(1989) to be published; see also Bul. Am. Phys. Soc. <u>33</u>, K 22-5 (1988)
21. D. W. Brown, K. Lindenberg and B. J. West, Phys. Rev. B <u>37</u>, 2946 (1988)
22. T. S. Rahman, R. S. Knox and V. M. Kenkre, Chem. Phys. (1978)
23. V. M. Kenkre and H. -L. Wu, Bul. Am. Phys. Soc.<u>33</u>, no. K 22-6 (1988)
24. M. Toda, <u>Theory of Nonlinear Lattices</u> (Springer-Verlag, Berlin 1981)
25. V. M. Kenkre, H. -L. Wu, and I. Howard, Bul. Am.Phys. Soc. <u>33</u>, K 22-4 (1988)

Quasiparticle Motion on a Chain with Alternating Site Energies and Intersite Interactions

V.I. Kovanis and V.M. Kenkre

Department of Physics and Astronomy, University of New Mexico,
Albuquerque, NM 87131, USA

We present exact calculations of the amplitude and the probability self-propagator of a quasiparticle moving among the sites of an infinite linear chain with alternating nearest-neighbour matrix elements for transfer and alternating site energies. A general solution is given and extreme limits are studied.

In our study of quasiparticle propagation on lattices, we have recently reported[1] some simple but interesting solutions for the probability of site occupation in two different but related one-dimensional infinite systems (chains). Although near-trivial in their simplicity, the solutions appear to have relevance to a variety of fields including exciton motion in molecular crystals[2-4], electron transport in superlattices[5,6] the motion of excitations in ferroelectric materials[7,8], and muon spin relaxation[9,10]. In the following, we present these solutions as well as calculations for a combined model which forms a generalization of the previous models. Although not of direct application to the interplay of nonlinearity and disorder, our analysis has a bearing on the subject of systematic inhomogeneities (systematic disorder) which has been attracting increasing attention in recent years as for instance in the context of quantum wells and superlattices.

Consider a quasiparticle such as an electron or an excitation (e.g. a Frenkel exciton) moving among the sites of a chain via nearest-neighbour matrix elements. The matrix elements alternate with the values $V \pm \delta$. The site energies also alternate, their values being $\pm 2\Delta$. The Hamiltonian $\mathcal{H}$ is

$$\mathcal{H} = 2\Delta \sum_m (-1)^m |m\rangle\langle m|$$
$$+ V \sum_m (|m\rangle\langle m+1| + |m\rangle\langle m-1|) + \sum_m \delta (-1)^m (|m\rangle\langle m+1| - |m\rangle\langle m-1|) \tag{1}$$

Here $|m\rangle$ represents a state of the moving quasiparticle localized on a lattice site m,(e.g. a Wannier state), and the summations are from $m = -\infty$ to $m = \infty$. By expanding the time dependent state $|\Psi(t)\rangle$ of the system in terms of the states $|m\rangle$ through $|\Psi(t)\rangle = \sum_m c_m(t)|m\rangle$, the evolution equation of the time-dependent amplitudes $c_m(t)$ is obtained as

$$i\partial_t c_m(t) = 2\Delta(-1)^m c_m(t) + \delta(-1)^m \{c_{m+1}(t) - c_{m-1}(t)\} + V\{c_{m+1}(t) + c_{m-1}(t)\} \tag{2}$$

where ∂_t denotes the time derivative. Discrete Fourier transforms of the real-space amplitudes $c_m(t)$ give us the following equations for the Bloch-space amplitudes $c^q(t)$:

$$i\partial_t c^q(t) = 2\Delta c^{q+\pi}(t) + 2i(\sin q)c^{q+\pi}(t) + 2V(\cos q)c^q(t) \tag{3}$$

$$i\partial_t c^{q+\pi}(t) = 2\Delta c^q(t) - 2i(\sin q)c^{q+\pi}(t) - 2V(\cos q)c^q(t) \tag{4}$$

Springer Proceedings in Physics, Vol. 39   **Disorder and Nonlinearity**
Editor: A.R. Bishop   © Springer-Verlag Berlin, Heidelberg 1989

We see that the dispersion relation between the energy Ω_q and the (dimensionless) wavevector q consists of the two branches

$$\Omega_q = \pm 2(\Delta^2 + \delta^2 \sin^2 q + V^2 \cos^2 q)^{\frac{1}{2}} \tag{5}$$

and the solution for the Bloch-space amplitudes for the initial condition that the quasiparticle occupies a single site (at m=0) is given by

$$c^q(t) = \cos(t\Omega_q) + 2i\{(\Delta + i\delta\sin q + V\cos q)/\Omega_q\}\sin(t\Omega_q) \tag{6}$$

We now invert (6) and evaluate the relevant integrals in the Laplace domain to obtain the explicit expression for the real-space amplitude propagator. For m = 0, the result is

$$\tilde{C}_0(\epsilon) = (\epsilon - 2i\Delta)\{(\epsilon^2 + \Omega_+^2)(\epsilon^2 + \Omega_-^2)\}^{-\frac{1}{2}} \tag{7}$$

which can be inverted explicitly to obtain the self-propagator in the time domain in terms of Bessel functions J_m and their convolutions. The frequencies $\Omega_\pm$ are given by

$$\Omega_+^2 = 4(\Delta^2 + V^2); \qquad\qquad \Omega_-^2 = 4(\Delta^2 + \delta^2) \tag{8}$$

Thus the amplitude self-propagator $C_0(t)$ and the probability self-propagator $\Psi_0(t)$ are given, respectively, by

$$C_0(t) = J_0(\Omega_+ t) + \Phi_1(t) + i\Phi_2(t) \tag{9}$$

$$\Psi_0(t) = J_0^2(\Omega_+ t) + \Phi_1^2(t) + \Phi_2^2(t) + 2J_0(\Omega_+ t)\Phi_1(t) \tag{10}$$

Equations (9) and (10) describe the motion of the initially site-localized quasiparticle, the frequency Ω_+ being given by (8) and the functions $\Phi_1(t)$ and $\Phi_2(t)$ by

$$\Phi_1(t) = \Omega_- \int_0^t d\tau J_0(\Omega_+ \tau) J_1(\Omega_-(t-\tau)) \tag{11}$$

$$\Phi_2(t) = 2\Delta \int_0^t d\tau J_0(\Omega_+ \tau) J_0(\Omega_-(t-\tau)) \tag{12}$$

By taking the limits $\Delta = 0$ and $\delta = 0$ it is straightforward to get the two limiting models of alternating site energies only, and of alternating intersite matrix elements as special cases. In the former, the Hamiltonian is

$$\mathcal{H} = 2\Delta\Sigma_m(-1)^m|m\rangle\langle m| + V\Sigma_m(|m\rangle\langle m+1| + |m\rangle\langle m-1|) \tag{13}$$

and in the latter it is

$$\mathcal{H} = \delta\Sigma_m(-1)^m(|m+1\rangle\langle m| - |m\rangle\langle m-1|) + V\Sigma_m(|m+1\rangle\langle m| + |m\rangle\langle m-1|) \tag{14}$$

Details of the motion in these two cases may be found in ref.[1].Here we briefly discuss only the physics. We will call the models of (13) and (14) respectively model A and model B.The effect of site energy variation (nonzero Δ) and that of intersite interaction variation (nonzero δ) are both that the quasiparticle slows down. When the chain disturbance is large,i.e $\Delta \gg V$ in model A and $2\delta^2 \gg V^2$ in model B,an interesting behaviour emerges.We see similarities as well as differences in the two models. Both show that the envelope of the self propagator follows the evolution of an undisturbed chain (i.e. with vanishing Δ and δ) but with an effective transfer matrix element V_{eff} which equals V^2/Δ for model A and $(V^2-\delta^2)/\delta$ for model B.

    The difference between the two models is made apparent by the fact that wild oscillations occur under the envelope for model B but not for model A. Indeed, the limit $\Delta \gg V$ reduces the oscillations in model A but the corresponding limit $2\delta^2 \gg V^2$ enhances them in model B. The oscillations describe the transfer of the quasiparticle within the two-site "unit cell" whose repetition generates chain B. The relative largeness of the matrix element V_1 which connects one site to the other in the unit cell (with respect to V_2 which connects the site to the next unit cell) causes the quasiparticle to remain largely within the cell and to move only a little from site to site. On the other hand, in model A, the quasiparticle encounters an equal energy mismatch for remaining in as well as leaving a unit cell.The sharp difference in the two models is seen as follows. When $V = \delta$ in model B, the connection between the cells is severed and the quasiparticle localized initially on the initial site oscillates (the self-propagator is simply the square of a cosine), being unable ever to leave the cell.We conclude by noticing that in the large Δ limit the dispersion relation (5) can be approximated and quantities such as the mean square displacement can be obtained.We also present fig.1 which describes the effective slowing down for model A $(\delta=0)$ and $(\Delta/V)=2$.

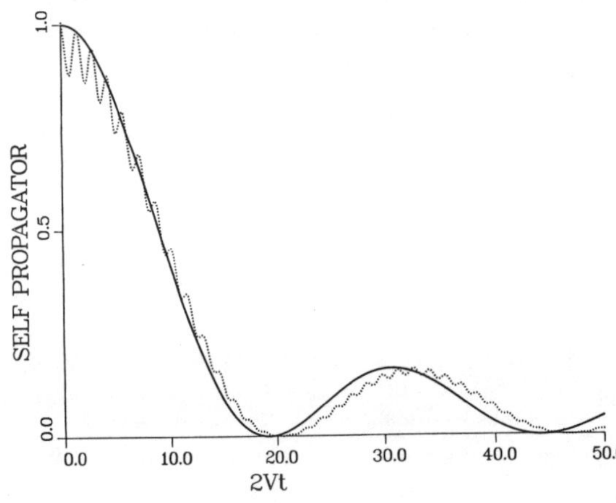

Fig.1 The time evolution of the probability self-propagator in model A for large energy mismatch compared to that for the degenerate chain with an effective intersite transfer. Shown is the exact evolution for model A (solid line) with $\Delta/V = 2$ and an undisturbed chain result viz. $J^2_0(2V_{eff}t)$ with $V_{eff} = V^2/\Delta$

References

1. V.I.Kovanis and V.M.Kenkre, Phys.Lett. A**130**, 147(1988)
2. J.P.Gallinar and D.C. Mattis, J.Phys. A**18**, 2583 (1985)

3.   V.M.Kenkre and P.Reineker, <u>Exciton Dynamics in Molecular Crystals and Aggregates,</u> ed.G.Holer (Springer,Berlin,1982).

4.   O.Bilek and Skala, Phys.Lett. A $\underline{119}$, 300(1986)

5.   R.Tsu and G.Dohler, Phys.Rev $\underline{12}$, 680(1975)

6.   P.Robin and M.W.Muller, Phys.Rev B $\underline{32}$, 5222(1986)

7.   J.S.Patel and D.M.Hanson,Chem.Phys.Lett. $\underline{89}$,473(1982)

8.   D.M.Hanson,Mol.Cryst.Liq.Cryst.$\underline{57}$, 243(1980)

9.   K.W. Kehr,K.Kitahara and K.Okamoura, J.Phys. Soc.Japan, $\underline{56}$, 428(1987)

10. K.W. Kehr and K.Kitahara ,J.Phys. Soc.Japan, $\underline{56}$, 889(1987)

Quasiperiodic Physics in One Dimension*

B. Sutherland

Department of Physics, University of Utah, Salt Lake City, UT 84112, USA

In my talk today, I wish to make two points. First, there are natural quasiperiodic sequences of two elements, which we call A and B, for any ratio of B's to A's, either rational or irrational. These sequences generalize the well-known Fibonacci lattice based on the golden mean, and are natural in the sense that they arise in many physical situations, independently of the exact nature of the interactions. Second, for these natural sequences most of the techniques based on exact scaling relations which have been developed for treating the Fibonacci lattice [1,2,3], and which have proven so powerful, can be used.

1. THE SEQUENCES

Let us first define the sequences that we will study, and briefly exhibit their important properties [4,5]. We will introduce a geometric picture; all of our remarks are illustrated in Fig. 1. Consider a piece of x-y graph paper with grid lines drawn only at integer x -- the B-lines -- or integer y -- the A-lines. We then draw the straight line L: $y = \lambda x + \phi$. We will assume that $\lambda \geq 1$, since otherwise we could exchange A and B. Further, we may as well take ϕ such that $1 > \phi \geq 0$. This line cuts the A and B lines in order, determining the cutting sequence of A's and B's. This cutting sequence can more easily be seen by drawing a broken line L' which connects the centers of each cell of the graph paper as it is visited by the line L. The cutting sequence is the same for the two lines L and L', and L' is a "best" staircase approximation to the slope λ. Line L' never wanders further than $1/\sqrt{2}$ from line L. As we walk along the line L, we find the A cuts are spaced a distance $\alpha = \sqrt{\lambda^2+1}/\lambda$ apart, while the B cuts are spaced a distance $\beta = \sqrt{\lambda^2+1}$ apart. Since $\lambda \geq 1$, $\beta \geq \alpha$. On the other hand, as we walk along the broken line L', all cuts are spaced a unit distance apart, and L' is our physical lattice. The exact physical nature of the A's and B's will be specified later. On both lines L and L', let us place the origin at the B-cut at x=0. On the average, as we move a distance $\alpha+\beta$ on L', we move a distance $\sqrt{\alpha^2+\beta^2}$ on L, so if we draw L and L' so that the ratio of the scales is $(1+\lambda)/\sqrt{\lambda^2+1}$, the two lattices will stay in register on the average. All of this is shown in Fig. 1.

The first remark is that since we cannot fit a β within an α, then between any two B's on L, there is at least one A, and so the B's are not isolated. Let $[\lambda]$ be the greatest integer less than or equal to λ, or equivalently the integer part of λ. Since $[\lambda] \leq \lambda = \beta/\alpha < [\lambda]+1$, then $\alpha[\lambda] \leq \beta < \alpha([\lambda]+1)$. Therefore, we can fit $[\lambda]$ α's within a β, but

---

* Work supported in part by National Science Foundation grant #DMR 86-15609.

Springer Proceedings in Physics, Vol. 39   **Disorder and Nonlinearity**
Editor: A.R. Bishop   © Springer-Verlag Berlin, Heidelberg 1989

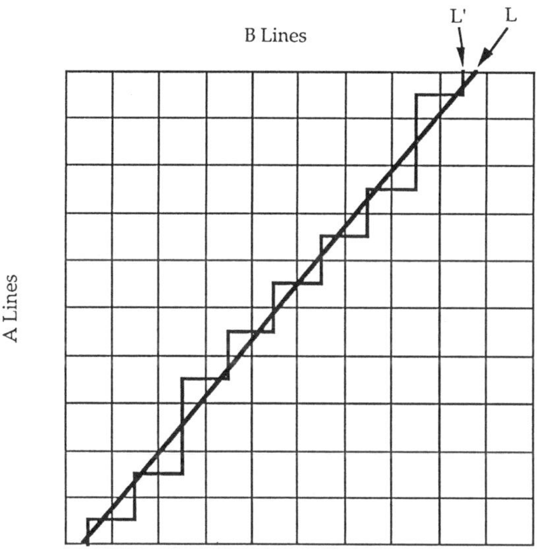

Figure 1   The explanation of this figure is given in the text.

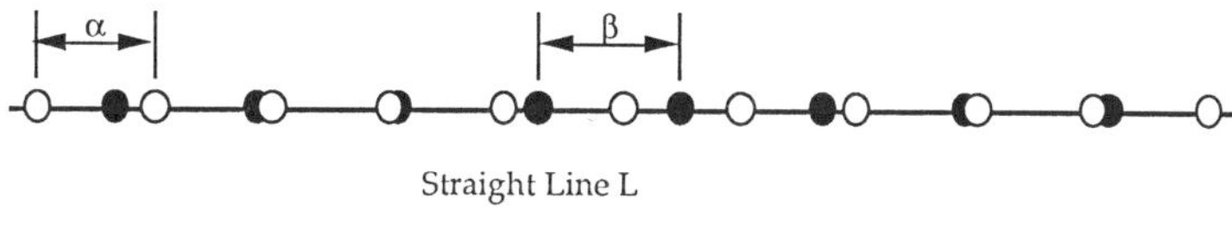

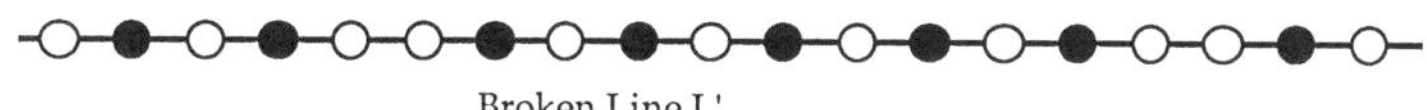

not $[\lambda]+1$ α's, and so between neighboring B's, there are either $[\lambda]$ or $[\lambda]+1$ A's. Thus, on L', the nearest-neighbor spacing of B's is either $[\lambda]+1$ or $[\lambda]+2$.

However, knowing this, the fraction of each is determined. The average nearest-neighbor spacing d of B's on L' is $d = 1/(\text{fraction of B's})$. Since on the average L and L' have the same slope $\lambda = (\text{number of A's})/(\text{number of B's})$, then $1+\lambda=1/(\text{fraction of B's})$, and so $d = 1+\lambda$. Therefore if f_1 is the fraction of nearest-neighbor spacings on L' of size $[\lambda]+2$, and $1-f_1$ is the fraction of nearest-neighbor spacings of size $[\lambda]+1$, then we have the sum rule

$$f_1([\lambda]+2) + (1-f_1)([\lambda]+1) = [\lambda]+1+f_1 = 1+\lambda \ \text{ or } \ \lambda = [\lambda]+f_1.$$

Therefore, f_1 is the fractional part of λ.

This can be continued. Between any two k^{th} neighbor B's there are either m_k or m_k-1 A's, where $\alpha(m_k-1)\leq\beta_k<\alpha m_k$. Again, a sum rule gives the fraction f_k of kth neighbor spacings of size $m_k +1$ by

$$f_k(m_k+1) + (1-f_k)m_k = k(\lambda+1) \quad \text{or} \quad m_k+f_k = k(\lambda+1) \quad \text{or} \quad m_k = [kd].$$

Thus m_k and f_k are the integer and fractional parts of $k(\lambda+1) = kd$ respectively.

As we walk along L, the B's are located at $j\beta$, while the A's are located at $k\alpha-\psi$, where $\alpha>\psi=\phi\sqrt{\lambda^2+1}/\lambda\geq 0$. Here j and k are integers. Then the number of A's between the B at 0 and the B at $j\beta$, is given by finding the largest k such that

$$k\alpha - \psi < j\beta \quad \text{or} \quad k < j\lambda + \psi/\alpha \quad \text{or} \quad k = [j\lambda + \psi/\alpha] \text{ , with } 1>\psi/\alpha\geq 0.$$

Let x_j be the integer location of the jth B on the lattice L'. Then we have established that $x_j = j + k = j + [j\lambda+\psi/\alpha] = [jd+\psi/\alpha]$. This is very reminiscent of the previous formulas for the kth neighbor spacings.

Let $n_0 = [\lambda]$ or $n_0+1 = [\lambda]+1$ be the number of A's which fit between two neighboring B's. As we saw, n_0+1 occurs with frequency $\delta_0 = f_1$, and n_0 occurs with frequency $1-\delta_0 = 1-f_1$, where $n_0 + \delta_0 = \lambda$ is a decomposition of λ into integer and fractional parts, so that $n_0 = [\lambda]$ and $\delta_0 = \lambda-[\lambda]$.

Now suppose on L' we make the replacements

$$A^{n_0}B \rightarrow A \quad \text{and} \quad A^{n_0+1}B \rightarrow BA,$$

or equivalently,

$$A \rightarrow B \quad \text{and} \quad A^{n_0}B \rightarrow A.$$

We now have deflated our original lattice L' to a new lattice L", where the ratio of B's to A's is now $\delta_0 = \lambda-[\lambda]$. Defining a new λ_1 as the ratio of A's to B's on L", we have $\lambda_1 = 1/\delta_0$. This process can be continued indefinitely through a succession of deflated lattices $L^{(k)}$ by the deflations

$$A^{n_k}B \rightarrow A \quad \text{and} \quad A^{n_k+1}B \rightarrow AB, \text{ or } A \rightarrow B \text{ and } A^{n_k}B \rightarrow A.$$

and the integers n_k are just the integers in the continued fraction expansion of λ:

$$\lambda = n_0 + \cfrac{1}{n_1 + \cfrac{1}{n_2 + ...}} \equiv [n_0,n_1,n_2,...].$$

This then gives the scaling transformations for these natural sequences.

2. PHYSICAL MOTIVATION FOR THE SEQUENCES

Let us suppose that the A's and B's represent two types of particles that occupy the sites of the one-dimensional lattice L'; in fact, without loss of generality, let A be a hole or vacancy, and B be a particle. The average spacing between B's is d = 1/density, and we assume as before that $d\geq 2$. Further, suppose that the B's a distance r apart interact with a potential V(r). We claim that if V(r) satisfies the condition

$$V(a) + V(b) < V(a+1) + V(b-1), \quad \text{for } N > a \geq b > 0,$$

then the sequences of the previous section minimize the energy. A repulsive continuous potential which falls to zero at infinity, for instance, would satisfy the condition. (The upper limit N is necessary only if we have to approximate the infinite lattice through a sequence of finite periodic lattices of period N.) This is a result first shown by Hubbard [6], and by Pokrovsky and Uimin [7].

We see the result as follows: Let us first examine the energy of nearest-neighbor interactions only. We label the nearest-neighbor separations as d_j. Then the nearest-neighbor energy is $\Sigma V(d_j)$. But if d_{max} and d_{min} denote the maximum and minimum values of d_j respectively, then for potentials which satisfy our condition, we can always lower this energy by taking instead $d_{max}-1$ and $d_{min}+1$, <u>unless</u> $d_{max}-d_{min} = 1$. Thus, an absolute minimum of the energy is obtained if the kth neighbor spacings take only two values which differ by one. However, as we saw in the previous section, the cutting sequences have precisely this property. And by construction, they are unique.

Thus, if there is repulsion between like B particles of the lattice gas, then we have established that for a given density $1/d$ of B's, the equilibrium distribution will be with B's located at sites $x_j = [jd]$, or translates thereof. These are the configurations that will be annealed in no matter what the detailed nature of the repulsive potential, and for this reason we call the sequences "natural".

3. FINITE APPROXIMATIONS

We will approximate the infinite lattice L' through a sequence of finite lattices L_k obtained by inflation from a "seed" A. Thus, let the inflation transformation T_n be given by

$$T_n (A,B) = (A^nB,A).$$

This is the inverse of the previous deflation transformation.

We then approximate λ by truncating the continued fraction expansion at level k, so that

$$\lambda \cong n_0 + \cfrac{1}{n_1 + \cfrac{1}{n_2 + \cdots \cfrac{}{n_k}}} \equiv [n_0,n_1,n_2,...,n_k].$$

Then L_k is constructed by successive inflation, so

$$(L_{k+1},L_k) = T_{n_0} \cdots T_{n_k} (A,B).$$

The first few lattices are

$$L_{-1} = B,$$
$$L_0 = A,$$
$$L_1 = A^{n_0}B = L_0^{n_0}L_{-1},$$
$$L_2 = (A^{n_0}B)^{n_1}A = L_1^{n_1}L_0,$$

$$L_3 = ((A^{n_0}B)^{n_1}A)^{n_2}A^{n_0}B = L_2^{n_2}L_1,$$
$$\ldots$$
$$L_{k+1} = L_k^{n_k}L_{k-1}.$$

(Here multiplication of lattices is understood as concatenation of symbols.) The final formula can be seen because after applying T_{n_k} we have T_{n_k} (A,B) = (A^{n_k}B,A), and so application of the remaining inflation operators just replaces (A,B) with (L_k, L_{k-1}). This formula can be iterated.

Likewise, if we let N_k denote the length of L_k, then we obtain the recursion relation,

$$N_{k+1} = n_k N_k + N_{k-1} , \text{ with } N_0 = N_{-1} = 1.$$

Iteration of this equation gives the bounds,

$$n_k n_{k-1}...n_1 n_0 F_{k+1} \leq N_{k+1} \leq n_k n_{k-1}...n_1 n_0 \text{ or } F_{k+1}.$$

Here F_k are the Fibonacci numbers, defined by the previous recursion relation with all n_k=1. Thus N_k grows at least as fast as the Fibonacci numbers, which grow exponentially.

4. TWO PHYSICAL EXAMPLES

In this talk, we shall illustrate the method with two physical examples. The first is a case where the quasiperiodic sequence is a sequence in time. Let us consider a spin-1/2 in a pulsed magnetic field. The pulses are of two types -- A or B -- according to the direction and strength of the magnetic field. The time evolution of the spin is governed by the time-dependent Schrodinger equation

$$i\frac{\partial \Psi}{\partial t} = -\mu \mathbf{B} \cdot \sigma \Psi,$$

with σ the Pauli spin operators. Thus the time evolution of Ψ due to a pulse of type A is given by a 2x2 unitary matrix, which we also call A, and which is given by

$$A = \exp[i\mu\sigma \cdot \int \mathbf{B}_A(t)\, dt].$$

Likewise there is a B matrix, and both A and B are elements of SU(2). The time evolution operator for the nth pulse we write as M(n), and it is either A or B.

The second example is a one-dimensional lattice, and it exhibits quasiperiodicity in space. We consider a tight-binding or hopping problem governed by the time-independent Schrodinger equation

$$-\psi_{n+1} - \psi_{n-1} + V_n \psi_n = E \psi_n.$$

If we iterate the equation, we have a two point iteration, and defining the vector $\Psi(n) = (\psi_n, \psi_{n-1})$, the iteration becomes $\Psi(n) = M(n) \Psi(n)$. The 2x2 matrix M(n) is the transfer matrix

$$M(n) = \begin{pmatrix} V_n - E & -1 \\ 1 & 0 \end{pmatrix}$$

and we assume that the on-site potential takes only the two values V_A or V_B.

Then $M(n)$ takes only two values, which we again call A and B. These matrices are elements of $SL(2,R)$.

Thus in both examples, we multiply matrices $M(n)$ together, which take only two values according to our natural sequences, and the net effect is in turn given by

$$D(n) = M(n) M(n-1) ... M(2) M(1).$$

In fact, let us only examine the total effect at the end of one of our finite lattices L_k, which is given by a single matrix $D_k = D(N_k)$. Then the matrices themselves obey the recursion relation

$$D_{k+1} = D_k^{n_k} D_{k-1}.$$

The matrices D are elements of a matrix group G, and the recursion relation on pairs (D_k, D_{k-1}) determines a dynamics on GxG. This is an exact scaling relation, and allows us to leap-frog to exponentially larger and larger matrices.

5. SCALING TRANSFORMATIONS

As we have seen, the inflation-deflation transformations are of the form

$$T_n (A,B) = (A^n B, A).$$

These transformations however can be generated by the following two elementary transformations:

$R (A,B) = (BA,B)$, left multiplication, and $X (A,B) = (B,A)$, exchange.

For instance $T_n = R^n X$.

Then assuming that A and B are elements of a matrix group G, we define

$$\Lambda = \mathrm{trace}(A^{-1}B^{-1}AB) = \mathrm{trace}(ABA^{-1}B^{-1}).$$

We make one further assumption that there exists a matrix U such that $A^{-1} = UAU^{-1}$ for all A in G. Then $\mathrm{trace}(A^{-1}) = \mathrm{trace}(A)$, and so

$$\Lambda = \mathrm{trace}(B^{-1}A^{-1}BA).$$

One easily sees that A is invariant under X and R:

$$X\Lambda = \mathrm{trace}(B^{-1}A^{-1}BA) = \Lambda, \text{ and}$$
$$R\Lambda = \mathrm{trace}(A^{-1}B^{-1}B^{-1}BAB) = \mathrm{trace}(A^{-1}B^{-1}AB) = \Lambda.$$

Thus X and R generate a group S which acts on GxG, and includes the inflation-deflation transformations T_n, and Λ is an invariant of S. The exact structure of S, however, is unclear to us at the moment.

However the group S commutes with the inner automorphisms of G, since if g is any element of G, and t is an element of S, then

$$t[g(A,B)g^{-1}] = g[t(A,B)]g^{-1}.$$

This implies that S induces a symmetry group S', called the trace map, on the invariants of (A,B) under inner automorphisms. Further, Λ must be a function only of the invariants of (A,B) under inner automorphisms, and is also an invariant of S'.

We now restrict ourselves to G = SL(2,C), which includes the two physical examples of section 4. Then there are only three independent invariants of (A,B), which we take to be

$$(x,y,z) = (\text{trace}(B)/2, \text{trace}(A)/2, \text{trace}(AB)/2 = \text{trace}(BA)/2).$$

(We note that for SL(2,C), our assumption that trace(g) = trace(g^{-1}) is satisfied.)

One easily shows that for SL(2,C),

$$\Lambda = 4(x^2+y^2+z^2) - 8xyz - 2.$$

First of all, Λ is invariant under permutations of x,y,z, and invariant if we change the signs of any pair of x,y,z. Thus Λ has the symmetry of a tetrahedron. We might now examine the group S' induced by S on the invariant surface λ of (x,y,z). X itself is just a permutation and so is included in the tetrahedron group. However, R is more complicated, and thus we will postpone a more full discussion of the trace map until a longer paper.

6. THE 6-CYCLE OF THE TRACE MAP

Instead considering the trace dynamics in general, let us look at a particular and important invariant set of the trace map -- the so-called 6-cycle:

$$(a,0,0) = \psi_1, \quad (-a,0,0) = -\psi_1,$$
$$(0,a,0) = \psi_2, \quad (0,-a,0) = -\psi_2,$$
$$(0,0,a) = \psi_3, \quad (0,0,-a) = -\psi_3.$$

Here, a is related to the invariant Λ by $a = \sqrt{\Lambda+2}/2$.

Now we determine the representation of the generators X, R on the invariant 6-cycle. First, we see that the action of the exchange X is

$$X\psi_1 = \psi_2, \quad X\psi_2 = \psi_1, \quad X\psi_3 = \psi_3,$$

and so the matrix representation of X is

$$X = \begin{pmatrix} 0 & 1 & 0 \\ 1 & 0 & 0 \\ 0 & 0 & 1 \end{pmatrix} = X^t, \quad \det X = -1.$$

Likewise, for the left multiplication R, the action is

$$R \psi_1 = (a,0,0) = \psi_1, \quad R \psi_2 = (0,0,-a) = -\psi_3, \quad R \psi_3 = (0,a,0) = \psi_2,$$

and so the matrix representation of R is

$$R = \begin{pmatrix} 1 & 0 & 0 \\ 0 & 0 & 1 \\ 0 & -1 & 0 \end{pmatrix}, \quad R^{-1} = R^t, \quad \det R = 1.$$

Now it is easy to see that this group is the full group of symmetries of the octahedron, including both reflections and rotations. The order of this group is 48. X is a reflection about the x=y plane, and R is a rotation about the x-axis by $\pi/2$. Therefore, the trace map hops about from one group element to another, with a time dependent dynamics determined by the continued fraction expansion of λ -- the ratio of A's to B's in the sequence.

If λ is the golden mean -- or indeed for λ any quadratic irrational -- the continued fraction expansion is periodic, and hence since the phase space is finite, and all elements must be fixed points if we iterate a sufficient number of times. For other irrationalities, the situation is less clear, but currently under investigation.

This concludes the talk, and to summarize: We have shown that many of the properties of the one dimensional Fibonacci lattice based on the golden mean are shared by a more general lattice with arbitrary and in general incommensurate density and sublattice spacing. Much of this work was done in collaboration with K. Iguchi.

7. REFERENCES

1. M. Kohmoto, L. P. Kadanoff and C. Tang, Phys. Rev. Lett. **50**, 1870 (1983).

2. S. Ostlund, R. Pandit, D. Rand, H. J. Schellnhuber and E. Siggia, Phys. Rev. Lett. **50**, 1873 (1983).

3. B. Sutherland and M. Kohmoto, Phys. Rev. B **36**, 5877 (1987).

4. C. Series, Math. Intelligencer **7**, 20 (1985).

5. H. Davenport, *The Higher Arithmetic* (Cambridge, Cambridge, 1982).

6. J. Hubbard, Phys. Rev. B **17**, 494 (1978).

7. V. L. Pokrovsky and G, V. Uimin, J. Phys. C **11**, 3535 (1978).

Density of States in Disordered Two-Dimensional Electron Systems

Bing C. Xu

Department of Chemistry, Brown University, Providence, RI 02912, USA

Abstract By using a supersymmetric mean-field formulation, the density of states of a disordered two-dimensional electron system is calculated, which is quite different from previous results obtained either by using single impurity approximation or by neglecting couplings between Landau levels, and qualitatively agrees with experimental results.

For systems having large number of random impurities, defects etc., one needs to ensemble average over the disorder to achieve answers as long as the statistical distribution laws of the random disorder are known. Treating random disorder problems, supersymmetry method [1],[2],[3] proves to be very useful. Its mathematical elegancy, simple but rigorous procedures overcome the difficulties met by method like Replica[4]. For systems with Gaussian, Poisson distributed random disorder one may successfully use supersymmetric methods to get nontrivial results. In general, by taking advantages of both Grassman and conventional integrals (supersymmetric integrals) one is able to average over the disorder potential exactly at an early stage in calculations and arrive at an effective quantum field theory with supersymmetric lagrangian. Couple of one-dimensional disordered problems are solved exactly using this method, see Refs. [1], [3], however, for higher dimensional problems, there exists no general exact solution to author's knowledge. The major difficulty is the nonlinearity of the effective Lagrangian obtained after averaging over the disorder, thus the simple transfer operator method is no longer valid for these higher (> 1) dimensional problems.

In this paper, we use the supersymmetric functional integral formalism to average over random disorder at very beginning of computing averaged Greens function of a two-dimensional disordered electron system with a magnetic field applied along $\hat{z}$-direction, hence convert the original disordered Lagrangian to an effective supersymmetric Lagrangian involving superfields which are similar to the usual quantum field Φ^4 Lagrangian without disorder, we therefore can apply standard approximation schemes (tree level approximation in this case) in quantum field theory to solve the problem.

Springer Proceedings in Physics, Vol. 39   **Disorder and Nonlinearity**
Editor: A.R. Bishop   © Springer-Verlag Berlin, Heidelberg 1989

Using the same SUSY technique, Brezin et al.[2] were able to solve a similar problem but with an assumption that neglecting Landau levels coupling, i.e., assuming that those Landau levels are well separated from each other. The resulting density of states, surely, are also well separated without overlaping. However, recent experimental results[5],[6] have shown that those Landau levels are actually overlaped pretty much which implies that the above assumption is inadequate. The density of states obtained by using so called self-consistent Born approximation[7] which is, basicly, a single impurity approximation, is similar to that of Brezin et al. The treatment we adopted is a simple supersymmetric mean-field approximation, but it automatically includes all other Landau levels contributions in a self-consistent manner. The resulting density of states is quite different from those of previous investigation, yet, qualitativly agrees with experimental results.

In general, for Gaussian distributed random disorder we have $< U(\mathbf{r})U(\mathbf{r}') >= \gamma\delta(\mathbf{r} - \mathbf{r}')$, where the angular bracket denotes the average over the random realization, γ is a constant representing the strength of the disorder. For charged impurity, one may model the disorder to be non-zero mean Gaussian distributed, i.e., $< U(\mathbf{r}) >= U_0$, and for δ-function type weak impurity, one can prove that $U_0 = cv, \quad \gamma = cv^2$ where c is the concentration of the impurities and v is the scattering strength of the impurity.

Writing explicitly the Hamiltonian of the system (without considering the electonic spin degrees of freedom as usual), we have

$$\hat{H} = \frac{1}{2m}(\hat{\mathbf{P}} - \frac{e}{c}\mathbf{A})^2 + U(\mathbf{r}) . \tag{1}$$

Although the spectrum of the Hamiltonian can be found in the absence of disorder potential $U(\mathbf{r})$[11] (assuming Landau gauge), in the presence of $U(\mathbf{r})$ the spectrum can not be found. Fortunately, one is usually only interested in quantities which are averaged over the disorder. For example, the calculation of thermodynamic quantities requires only knowledge of the averaged density of states which is given by the following expression[10]

$$< \rho(E) >= - \frac{1}{\pi}tr < ImG^R(\mathbf{r},\mathbf{r};E) >, \tag{2}$$

where G^R denotes the retarded Green function which can be formally written in terms of the (unknown) eigenfunctions $\{\phi_k\}$ and eigenvalues $\{E_k\}$ of the full Hamiltonian as

$$G^R(\mathbf{r},\mathbf{r}';E) = \sum_k \frac{\phi_k(\mathbf{r})\phi_k^*(\mathbf{r}')}{E - E_k + i\delta} = \sum_k \phi_k(\mathbf{r})\phi_k^*(\mathbf{r}')G^R(k;E) . \tag{3}$$

By using Grassmannian integration, we can also write the Green function as

$$G(\mathbf{r},\mathbf{r}';E) = -iZ^{-1} \int D\phi D\phi^* \, \phi\phi^* \, exp[- \int L_\phi dr], \tag{4}$$

where

$$Z^{-1} = \int D\psi D\psi^* exp[-\int L_\psi d\mathbf{r}], \tag{5}$$

or equivalently

$$G(\mathbf{r}, \mathbf{r}'; E) = -i \int D\phi D\phi^* D\psi D\psi^* \ \phi\phi^* exp[-\int L d\mathbf{r}], \tag{6}$$

where

$$L = i[\phi^\dagger(\epsilon - H - i\delta)\phi + \psi^\dagger(\epsilon - H - i\delta)\psi] = L_\phi + L_\psi \tag{7}$$

and ϕ, ψ are the bosonic and fermionic components of a superfield $\Phi = \begin{pmatrix} \phi \\ \psi \end{pmatrix}$.

Note that the supersymmetric Green function Eq.(6) has no denominator which conventional Green function does have is the central importance of this method. This makes it easier for us to average over the random disorder potential at an early stage of calculation without using any trick, like Replica[4] trick, or approximation.

Now the average over random potentials can be done easily to yield

$$\mathcal{L} = i\Phi^\dagger[\frac{1}{2m}(\hat{\mathbf{P}} - \frac{e}{c}\mathbf{A})^2 + E - i\delta]\Phi + (\gamma/2)(\Phi^*\Phi)^2. \tag{8}$$

Eq. (8) is a typical Lagrangian of supersymmetric ϕ^4 model in field theory. Using Replica trick one can also get the same average , but instead, it includes n-replica fields, also one has to take the replica index $n \to 0$ at the end which is mathematically unjustfiable, thus one can not be sure whether the replica results are correct if there is no independent check, see Ref. [8] for example where using replica trick yields incorrect results.

The averaged density of electronic states is obtained from the imaginary part of the Green function according to Eq. (2). For 2-dimensional problem (as well as 3-dimensional), one can not use the simple transfer operator method such as employed in references [1] and [3] where exact results can be obtained. Here we are left with a coordinate-dependent (no simple translational invariance) quantum superfield problem and thus appropriate approximation schemes must be devised to perform the functional integrations over the superfields.

As a reasonable start one can use a mean field approximation to handle the quartic (nonlinear) term in the Lagrangian, i.e., $(\Phi^*\Phi)^2$, quadratizing the Lagrangian as follows:

$$\mathcal{L} = i\Phi^\dagger[\frac{1}{2m}(\hat{\mathbf{P}} - \frac{e}{c}\mathbf{A})^2 + E - i\delta]\Phi + \gamma(\Phi^*\Phi) < \Phi\Phi^* > \tag{9}$$

where the field operator appearing in between the angular brackets is now a number operator. For this quadratic Lagrangian, the functional integration is easily performed to yeild

$$tr < G > \; = \; -i/(\pi) \sum_n \frac{1}{i[-E_n + E - i\delta] + \gamma < \Phi\Phi^* >}$$
$$+ \; -\frac{i}{\pi} \int d\vec{p} \, \frac{1}{i[E - p^2/2m - i\delta] + \gamma < \Phi\Phi^* >}, \qquad (10)$$

where the summation represents the discrete spectrum and the integration represents the continuum spectrum, and E_n is the pure energy spectrum of 2-D electron system in a magnetic field applied along $\hat{z}$-direction without disorder potential.

By using the equation $itr < G >=< \Phi\Phi^* >$ and factoring out i in the denominator of (10) we obtain a self-consistent equation for the averaged Greens function (we are more interested in discrete part of the density of states)

$$tr < G >= -1/(\pi) \sum_n \frac{1}{[-E_n + E - i\delta] - \gamma tr < G >}. \qquad (11)$$

This is the main equation of the mean-field approximation. Solving for $tr < G >$ self-consistently, one can obtain the density of states which is shown in the figure.

We note that, first, for $\gamma \to 0$ we have $\rho(E) \to \rho_0(E)$ where the subscript 0 denotes the pure system, which is $\delta-$function typy DOS representing discrete Landau levels as expected. Second, we note that the DOS does not possess any singularity for finite E, which is consistant with a general theorem proven by Wegner saying that the DOS of a disordered system does not have singularities[15], this can be proven easily from Eq.(11) by examining both real and imaginary parts of the Greens function in our case.

Third thing we note is that the tail for each broadened peak is not known for its functinal form, i.e., whether it has algebraic form or exponential form. It is still an open question in 2-D localization problem (as well as 3-dimensional systems) , namely, that adding external potential, do we still have all the states localized and is the band tail exponential ? (see discussions in Refs.[12],[13]), or an even stronger issue, namely, that without external potential, do we have all the states localized for this 2-D system (see [14] for discussions) ? However, it seems that there is no debate for one dimensional systems. The usual way to find the localization length and the mobility edge using this formalism is to find out the $< GG >$ correlation function which involves more calculations and is out of the scope of this simple treatment.

The fourth thing we note is that our treatment automatically includes all other Landau levels contributions, which has been neglected by most previous treatments[2],[7]. Comparing those with most recent experimental results[5],[6], the agreements are far from satisfactory. Due to the difficulties in dealing with the disorder potential, most of the previous work (except Ref.[2]) used basicly the single impurity approximation or the self-consistent Born approximation which supposed to be valid only at weak impurity limit.

However, in our treatment, we average the disorder exactly, converting the disorder problem to a quantum field problem and then making systematic approximation, this procedure proves to be better than making expansion first, averaging disorder term by term and then resumming the series, as a matter of fact, this traditional way can never get exact results (except Lloyd model[16]) say, in one-dimensional disordered systems, while SUSY method does give exact results such as in Refs. [1],[3]. This is because that some of the terms in the series cannot be resummed in the traditional method.

It is straightforward to compute the DOS numerically by self-consistently computing the imaginary part of the $tr < G >$.

The thin solid line shown in Fig. shows that $\bar{\gamma} \to 0$ (clean limit) which reproduces the density of states of the pure system where sharp peaks represent discrete Landau levels, while for finite but smaller $\bar{\gamma}$ (weak disorder) the discrete peaks are still well separated with small broadening, which is shown as the dashed line, again reproduces previous self-consistent Born approximation results as expected. We use dimensionless disorder strength and energy parameter where $\bar{\gamma} = \gamma/\omega_0^2$ and $\bar{E} = E/\omega_0$, ω_0 is energy spacing of Landau level. The dotdashed line is for a bigger $\bar{\gamma}$, where the broadening is much bigger than that of SCBA's density of states and at certain value of $\bar{\gamma}$, those peaks start to overlap, and finally disappear, i.e., one obtains flat density of states shown as the thick solid line at even bigger $\bar{\gamma}$ indicating strong disorder limit. Our results qualitatively agree with the experimental results[5]. Automatically including the coupling between all Landau levels in this simple self-consistent mean-field treatment is the major improvement to the previous methods. Recently, Das Samar and Xie recalculated the density of states of this 2-D electronic system based on traditional SCBA methods by including ladder and ring

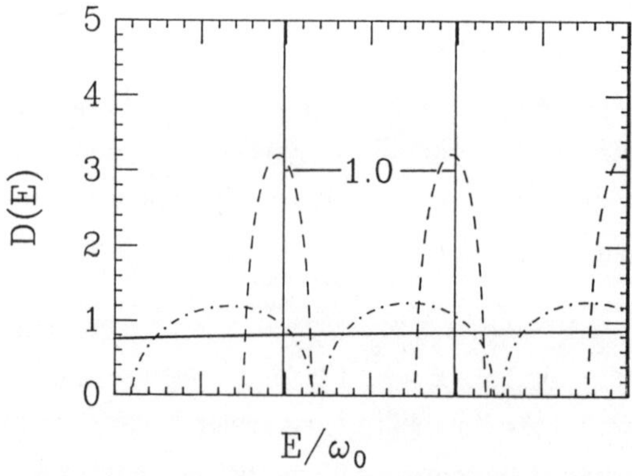

Density of states $D(E)$ vs. dimensionless energy parameter E/ω_0. Thin solid line, $\bar{\gamma} \to 0$; dashed line, $\bar{\gamma} = 0.005$; dotdashed line, $\bar{\gamma} = 0.04$; thick solid line, $\bar{\gamma} = 0.25$.

diagramms[17], the resulting DOS is quite similar to our simple mean-field resutl, however, more complicated calculations were involved in their treatment. By considering charged impurities case, one can treat the average again exactly with a additional term U_0 which will shift the Fermi level and the resulting DOS will depend on both parameter c-impurity concentration and v-impurity scattering strength.

The reason that this simple mean-field treatment can give reasonable results is that we have a relatively uniform physical background, i.e., that each Landau level broadening is similar and each unpurterbed Landau level is the same, there exist no dramatic fluctuations around those physical quantities in the system. However, for systems where physical quantities are not uniform, this simple mean-field treatment needs to be augmented by higher order corrections and fluctuations around the mean field, such as the case of a Hydrogenic liquid where each level broadening depends sensitively on spatial structures and disorder distributions, which is currently under investigation.

The author would like to thank R.M. Stratt and V. Dobrosavljevic for helpful discussions and the work was supported by the NSF under Grant No. CHE-8420214.

References

[1] K.B. Efetov, *Adv. Phys. 32, 53 (1983) and references therein.*

[2] E. Brezin, D.J. Gross and C. Itzykson, *Nucl. Phys. B235 [FS11] (1984).*

[3] B. C. Xu and S.E. Trullinger, *Phys. Rev. Lett. 57, 3113 (1986).*

[4] S.F. Edwards and P.W. Anderson, *J. Phys. F 5, 965(1975).*

[5] J.P. Eisenstein, H.L. Stormer, V. Narayanamurti, A.Y. Cho, A.C. Gossard, and C.W. Tu, *Phys. Rev. Letts. 55, 875, (1985).*

[6] F.F. Fang, T.P. Smith, and S.L. Wright, *Surf. Sci. 196, 310 (1988) and references therein.*

[7] T. Ando and Y. Murayama, *J. Phys. Soc. Jpn. 54, 1519 (1985); Phys. Rev. B 35, 2252 (1987).*

[8] J.J.M. Verbaarschot and M.R. Zimbauer, *Ann. Phys. (N.Y.) 158, 78 (1984).*

[9] B.C. Xu and R.M. Stratt, to be published.

[10] E.N. Economou, *Green's Functions in Quantum Physics*, (Springer-Verlag, Berlin, 1979).

[11] L.D. Landau and E.M. Lifshitz, *Theoretical Courses Physics, Vol. 3*, Pergamon Press (1977).

[12] H.L. Cycon, R.G.Froese, W. Kirsch and B. Simon, *Schrödinger Operators*, (Springer-Verlag, Berlin, 1987), Chp.9.

[13] *Phys. Today, Sep. 1988, p. 21 and references therein.*

[14] K.B. Efetov, *Sov. Phys. JEPT* 62 *(3) 605, (1985)*

[15] F. Wegner, *Z. Phys. B.* 44*, 9 (1981).*

[16] P. Lloyd, *J. Phys.* C2, *1717 (1969).*

[17] S. Das Sarma and X.C. Xie, *Phys. Rev. Letts.* 61*, 738, (1988).*

Part II

Solitons and Disorder

Soliton Scattering by Impurities.
An Analytical Approach to Interference Effects

Y.S. Kivshar, A.M. Kosevich, and A.O. Chubykalo

Institute for Low Temperature Physics and Engineering,
UkrSSR Academy of Sciences, 47 Lenin Avenue, SU-Kharkov 310164, USSR

We have studied analytically the soliton scattering by point and finite-size inhomogeneities on the basis of the perturbed sine-Gordon and Korteweg-de Vries equations. The most remarkable feature of such scattering is interference effects caused by the resonant emission of solitons. These effects may be described by a reflection coefficient which is calculated by means of soliton perturbation theory, the intensity of the inhomogeneities being the small perturbation parameter. Our results are applied to the theory of long Josephson junctions with installed inhomogeneities and to the theory of one-dimensional nonlinear atomic lattices with isotope-like point defects.

1. INTRODUCTION

In recent years great attention has been paid to the study of solitons as particle-like excitations of solids. Dynamical equations describing the evolution of soliton parameters in nearly integrable nonlinear systems are analogous to the equations of motion of classical particles. However, solitons are solutions of nonlinear field equations and, therefore, they should certainly have wave-like properties. In particular, the wave properties of solitons should appear during their scattering by impurities. Therefore, though soliton excitations are classical objects, in a number of problems they may demonstrate both corpuscular and wave-like properties.

The interaction of a soliton with an inhomogeneity leads to a non-stationarity in the soliton motion and is accompanied by emission of radiation in the form of linear waves. During the soliton scattering in one-dimensional systems the emission of linear waves in both forward and backward directions takes place. The interference phenomena is the most remarkable demonstration of wave properties of a soliton. The interference occurs when the ratio of the characteristic size of an inhomogeneity to the characteristic length of waves emitted by a soliton is half-integral. Therefore, as the simplest one-dimensional problem in which interference phenomena take place, we may consider either the system of two point impurities or one impurity with the length L. We consider these problems in the framework of the perturbed sine-Gordon and Korteweg-de Vries equations. These equations naturally arise in the theory of long Josephson junctions with installed inhomogeneities and in the theory of one-dimensional nonlinear lattices with isotope-like point impurities.

2. RESONANT EMISSION OF A FLUXON IN A LONG JOSEPHSON JUNCTION

2.1 Emitted Energy

In this section we deal with a real physical model which is described by the sine-Gordon (SG) equation, a long Josephson junction (JJ). A fluxon in a long JJ is known to correspond to solution of the kink type of the SG equation. It is known that

Springer Proceedings in Physics, Vol. 39   **Disorder and Nonlinearity**
Editor: A.R. Bishop   © Springer-Verlag Berlin, Heidelberg 1989

inhomogeneities can influence strongly the dynamics of fluxons. The fluxon may be captured by an inhomogeneity or escape from it at some values of bias current density, and the fluxon passing the inhomogeneity may generate an additional microwave radiation. These effects are observed as peculiarities of the junction current-voltage curves, the so-called zero-field steps (ZFSs) (see, e.g. PEDERSEN [1]).

The dynamics of magnetic flux nonlinear excitations in a long inhomogeneous JJ are known to be described by the perturbed SG equation

$$u_{tt} - [\varepsilon_1(x)u_x]_x + \varepsilon_2(x)sinu = f - \gamma u_t \,, \tag{1}$$

where standard variables and units are used (see, e.g. PEDERSEN [1]), u being the normalized magnetic flux density, and γ being the dissipation coefficient. To describe a JJ with a finite-size inhomogeneity we choose the functions $\varepsilon_1(x)$ and $\varepsilon_2(x)$ in Eq. (1) as follows

$$\varepsilon_n(x) = 1 + \varepsilon_n[\theta(x) - \theta(x - L)] \,, \quad n = 1, 2 \,, \tag{2}$$

$\theta(x)$ is the step function, L is the dimensionless length of the inhomogeneity. The typical values of L are 0.5 [GOLUBOV, USTINOV 2], 0.9-2.9 [VYSTAVKIN, DRACHEVSKY, KOSHELETZ, SERPUCHENKO 3] or 5 [AKOH, SAKAI, YAGI, HAYAKAWA 4]. We consider the case when f, γ, ε_1 and ε_2 in Eq. (1) are small parameters, so that we can use perturbation theory.

A fluxon in a homogeneous long JJ ($\varepsilon_1 = \varepsilon_2 = 0$) is known to be described by the expression

$$u_{f1} \approx f - 4tan^{-1}\big\{exp[(x - V_0t)/\sqrt{1 - V_0^2}]\big\} \,, \tag{3}$$

where $V_0 \approx [1 + (4\gamma/\pi f)^2]^{-1/2}$ is the velocity of the steady state fluxon motion.

When a fluxon moves near the inhomogeneity its velocity is changing and the fluxon generates the radiation in the form of linear waves. The spectral density $E(k)$ of energy emitted by a fluxon during its scattering by the inhomogeneity (2) may be obtained by means of the perturbation theory for solitons based on the inverse scattering transform [KIVSHAR, KOSEVICH, CHUBYKALO 5], [KIVSHAR, KOSEVICH, CHUBYKALO 6]

$$\begin{aligned}
E(k) =&(\pi/V^4)[\varepsilon_1 + \varepsilon_2(1 - V^2)]^2sin^2\Big[\frac{L}{2V}(kV - \sqrt{1 + k^2})\Big] \\
&\times sech^2[(\pi\sqrt{1 - V^2}/2V)\sqrt{1 + k^2}] \,.
\end{aligned} \tag{4}$$

The total emitted energy E_{cm} is determined by integrating of (4) over all wave numbers $|k| < \infty$. The dependence $E_{em}(V)$ is defined by numerical calculation of this integral. The characteristic feature of received function appears in its non-monotonic dependence on the fluxon velocity V (see Fig. 1), which characterizes a resonant emission: the waves emitted from two "edges" of a "box-like" inhomogeneity strengthen each other if the ratio of the time of the fluxon motion across an inhomogeneity $\tau = L/V$ to the period of generated waves $T = 2\pi/\omega', \omega'$ being the frequency in the fluxon reference frame ($\omega' \equiv \omega(k) - kV \equiv \sqrt{1 + k^2} - kV$), is half-integer [KIVSHAR, KOSEVICH, CHUBYKALO 5]. Resonant fluxon emission leads to the appearance of the additional maxima of $E_{em}(V)$ shown in Fig. 1.

Another interesting example of a physical system illustrating similar similar picture of radiative effects is a long JJ with two point inhomogeneities [KIVSHAR, KOSEVICH, CHUBYKALO 6]. In order to describe the interaction of a fluxon with two point inhomogeneities we have to choose $\varepsilon_n(x) = 1 + \varepsilon_n[\delta(x) + \delta(x - a)]$ in Eq. 1), a being the distance between the inhomogeneities. Simple calculations based on the inverse scattering transform gives the spectral density of the emitted energy (cf. (4))

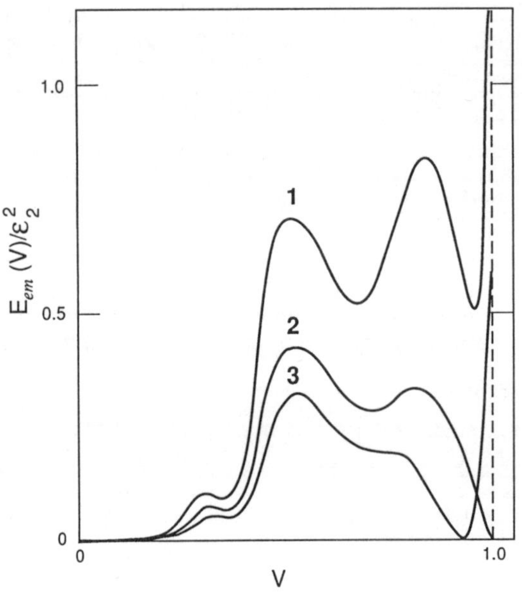

Fig. 1. The total emitted energy vs. the fluxon velocity for $L = 5$; (1) $\varepsilon_2/\varepsilon_1 = 0.2$, (2) $\varepsilon_2 = 0$, (3) $\varepsilon_2/\varepsilon_1 = -0.1$

$$\mathbf{E}(k) = (\pi/4V^6)[\varepsilon_1 + \varepsilon_2(1 - V^2)]^2 cos^2\left[\frac{2}{2V}(\omega - kV)\right]$$
$$\times (\omega - kV)^2 sech^2\left[(\pi\sqrt{1 - V^2}/2V)\sqrt{1 + k^2}\right] . \tag{5}$$

where $\omega \equiv \sqrt{1 + k^2}$. Having integrated this expression over all k, we obtain results similar to those for the case of a finite-size inhomogeneity, i.c. the appearance of additional maxima on the curve $E_{em}(V)$ due to the resonant fluxon emission.

The resonant fluxon emission may be characterized by the reflection coefficient R of a fluxon. We define this coefficient as a ratio of the energy $E_{em}^{(-)} \equiv \int_{-\infty}^0 \mathbf{E}(k)dk$ of waves emitted in the backward direction to the total fluxon energy $E_{fl} = 8(1 - V^2)^{-1/2}$. For example, using Eq. (5) one can find R as follows (for simplicity, $\varepsilon_1 = 0$)

$$R = \frac{\pi\varepsilon_2^2}{2^5V^6}(1 - V^2)^{5/2} \int_0^{\int} dk[\omega(k) + kV]^2$$
$$\times cos^2\left\{\frac{a}{2V}[\omega(k) + kV]\right\}sech^2[\pi\sqrt{1 - V^2}\omega(k)/2V] . \tag{6}$$

In the limiting case $a \to 0$ the reflection coefficient coincides with the value $4 \cdot R_1$, where R_1 is the soliton reflection coefficient for one point inhomogeneity. It is easy to obtain the asymptotics of the function $R/2R_1$ at small and large velocities (see KIVSHAR, KOSEVICH, CHUBYKALO [6]). For example, for the case $V << 1$ we have

$$R/2R_1 \approx 1 + [1 + (a/\pi)^2]^{-1/4}cos(a/V + const.) . \tag{7}$$

In the limit $a \to \infty$ the emission is non-resonant, and $R \approx 2R_1$. The function $R/2R_1$ is shown in Fig. 2 for $V = 0.4$ (curve 1) and $V = 0.9$ (curve 2). The form of the curve $R/2R_1$ depends essentially on the fluxon velocity and for large velocity has only a single minimum. The oscillating dependences appear as a result of the resonant fluxon scattering by inhomogeneities.

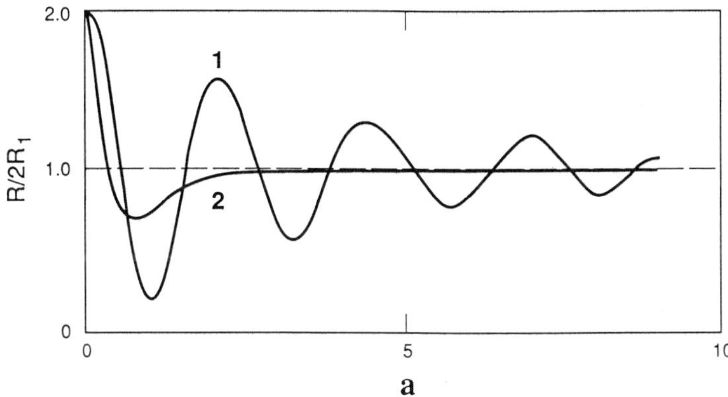

Fig. 2. The reflection coefficient of the fluxon vs. the distance between two point impurities, (1) $V = 0.4$, (2) $V = 0.9$

2.2 Zero-field Steps

One of the important characteristics of a JJ line is its current-voltage (IV) curve. It is this curve that may be obtained experimentally and gives us the important information about JJ's. Real long JJ lines have large but still finite length ℓ (e.g. $\ell \equiv 15$ in Refs. [GOLUBOV, USTINOV 2], [VYSTAVKIN, DRACHEVSKY, KOSHELETZ, SERPUCHENKO 3]). At zero magnetic field a fluxon oscillates between the junction's edges with the frequency $\sim V/\ell$. The IV-curve of such JJ is called the zero-field step (ZFS); for homogeneous JJ it has the form depicted on Fig. 3 by the dashed line.

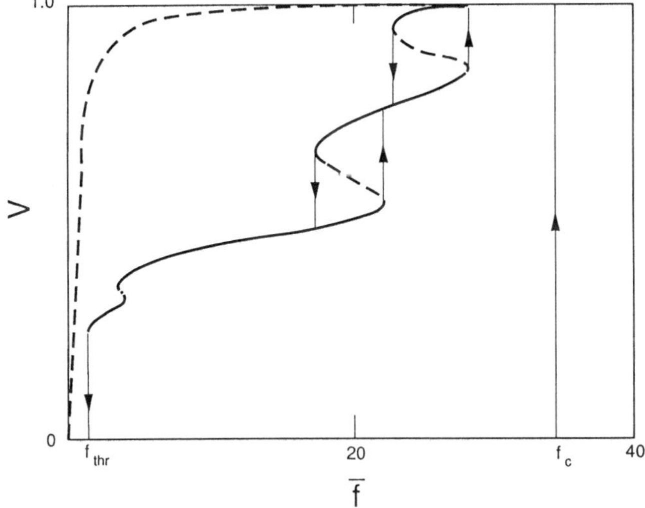

Fig. 3. The first zero-field step $V(\bar{f} = f/\gamma)$ of the inhomogeneous JJ with parameters $\bar{\varepsilon}_1 = \varepsilon_1/\sqrt{\ell\gamma} = 2$, $\bar{\varepsilon}_2 = \varepsilon_2/\sqrt{\ell\gamma} = 10$ and $L = 5$. The values f_c and f_{thr} are defined in [KIVSHAR, KOSEVICH, CHUBYKALO 5] and they correspond to capture and escape of a fluxon. The arrows show the direction of the motion along the curve with change of the external current $\sim f$.

Resonant fluxon scattering by an inhomogeneity can be observed in ZFS plots and leads to the appearance of additional peculiarities there. For a fluxon oscillating between JJ's edges, the continuous spectrum described above is transformed into a discrete one, consisting of many individual lines spaces $\sim \ell^{-1}$ apart. The envelope of this spectrum is close to the curve described by Eqs. (4) and (5). The total power P of this radiation can be expressed as follows:

$$P = 2E_{em}(V)/(\ell/V) \ .$$

Equating the power of the external force ($\sim f$) with the power of dissipative ($\sim \gamma$) and radiative ($\sim \varepsilon_1^2, \varepsilon_2^2$) losses, we obtain

$$f = \frac{4\gamma V}{\pi\sqrt{1-V^2}} + E_{em}(V)/\pi\ell \ . \tag{8}$$

The quantity f is proportional to the bias current density and the fluxon velocity, V, to the voltage in the transmission JJ line (see, e.g. PEDERSEN [1]). Therefore, Eq. (8) gives directly the ZFS of an inhomogeneous junction with a single trapped fluxon oscillating between the JJ's edges. The dependence $\bar{f} \equiv f/\gamma$ on V with $\bar{\varepsilon}_1 \equiv \varepsilon_1/\sqrt{\ell\gamma} = 2$, $\bar{\varepsilon}_2 \equiv \varepsilon_2/\sqrt{\ell\gamma} = 10$ for a finite-size inhomogeneity is shown in Fig. 3 at $L = 5$. The resonant fluxon emission appears in hysteresises.

3. LATTICE SOLITON SCATTERING IN NONLINEAR CHAINS

The monoatomic chain of first neighbor interacting particles in continuum approximation may be described by a generalized Boussinesq equation

$$u_{tt} - c_0^2 u_{xx} - p(u^2)_{xx} - q(u^3)_{xx} - h \cdot u_{xxxx} = 0 \ , \tag{9}$$

where $u = y_x$ and y denotes the longitudinal displacement of the atom with the co-ordinate x from its equilibrium position. The sense of other parameters in Eq. (9) is given in [FLYTZANIS, PNEVMATIKOS, REMOISSENET 7], and [LI, PNEVMATIKOS, ECONOMOU, SOUKOULIS 8]. We consider the inhomogeneous chain with two point-like isotopic defects, what should be described by the additional term in the r.h.s. of Eq. (1) $\varepsilon[\delta(x) + \delta(x-a)]u_{tt}$, $\varepsilon \equiv 1 - M/M_0$, M being the impurity mass. Making the standard assumptions for $h << 1$ we may obtain the perturbed modified KdV equation (see details in Ref. [KIVSHAR, KOSEVICH, CHUBYKALO 6])

$$w_\tau + p_0 w w_\xi + q_0 w^2 w_\xi + w_{\xi\xi\xi} =$$
$$= -(\varepsilon/2)[\delta(\tau + (h/2c_0^2)\xi) + \delta(\tau + (h/2c_0^2)\xi - \tau_0)]w_\xi \ , \tag{10}$$

where $w = u_\xi/A$, $\tau = (h/2c_0)t$, $\xi = x - c_0 t$, $p_0 = 2pA/h$, $q_0 = 2qA^2/2h$ and $\tau_0 \equiv ha/2c_0^2$. The parameter A represents the amplitude of the excitation. Considering ε as a small parameter we may use perturbation theory for the modified KdV equation. Let us consider, for simplicity, the case of the perturbed KdV equation, $p_0 = -6$, $q_0 = 0$. Then the soliton of the unperturbed equation has the form

$$w_s = -2\kappa^2 sech^2[\kappa(\xi - 4\kappa^2 t)] \ . \tag{11}$$

Calculating the reflection coefficient for the soliton (11) scattering from the point-like moving inhomogeneities from the r.h.s. of Eq. (10), one can obtain (for more details see Ref. [KIVSHAR, KOSEVICH, CHUBYKALO 6]):

$$R = (2^7\pi/3)\varepsilon^2(\kappa^3 h/c_0^2)^2 \int_0^\infty dz z^4 (1+z^2)^2$$
$$\times \cos^2[4\tau_0\kappa^3 z(1+z^2)]/sinh^2(\pi z) \ . \tag{12}$$

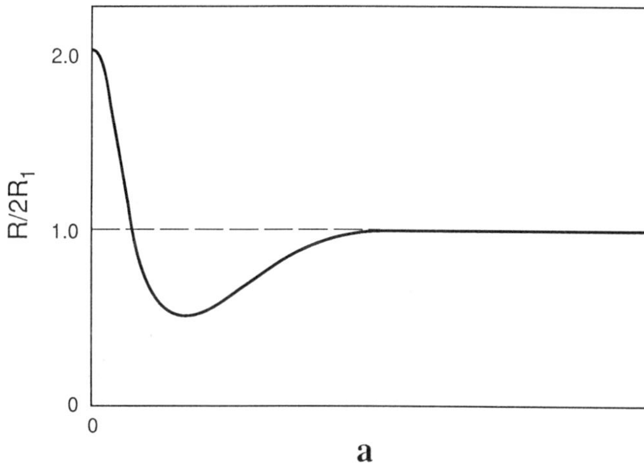

Fig. 4. The reflection coefficient of the KdV soliton vs. the distance between two point impurities, see Eq. (10).

The function (12) for $\tau_0 >> 1$ has the asymptotic form

$$R \approx 2R_1 - (16\sqrt{2\pi}\pi/3^{21/4})\varepsilon^2\alpha^2\frac{exp(-4B/3\sqrt{3})}{sin^2(\pi/\sqrt{3})} , \tag{13}$$

where $\alpha \equiv (2h/c_0^2)\kappa^3$, $B \equiv \alpha a$, and R_1 is the reflection coefficient for the one-impurity scattering. The function $R/2R_1$ is shown in Fig. 4. It is interesting that, unlike the case of the SG fluxon, this function is non-oscillating, i.e. the KdV soliton scattering is non-resonant. For $a >> 1$ the reflection coefficient tends to the value $2R_1$ (cf. (7)).

These results have the simple physical interpretation. In the case of the SG equation the energy spectral density $\mathbf{E}(k)$ concentrates for $V << 1$ in the region $|k| \leq 1$, i.e., in the scale of the frequency of emitted waves, in the region $\omega \leq 1$ (emitted waves have the frequencies which are close to the gap of the linear excitations). Therefore, in this case the emission from two impurities is resonant and the dependence $R(a)$ is oscillating, $\omega \equiv 1$ being the resonant frequency. In another case, when $V \leq 1$, the energy spectrum of emitted waves is rather wide (its width $\Delta k \sim (1-V^2)^{-1/2} >> 1$) and, as a result, the oscillations disappear. The case of the KdV equation is close to the "relativistic" $(1-V^2 << 1)$ SG fluxon scattering because the spectrum of the emitted energy of linear waves is wide. The detailed discussion of features of these problems may be found in [KIVSHAR, KOSEVICH, CHUBYKALO 6] where a qualitative comparison with the numerical results of [FLYTZANIS, PNEVMATIKOS, REMOISSENET 7], and [PNEVMATIKOS, ECONOMOU, SOUKOULIS 8] is also presented.

REFERENCES

1. N. F. Pedersen: In **Solitons**, eds. by S. E. Trullinger, V. E. Zakharov and V. L. Pokrovsky (North Holland, Amsterdam, 1986) p. 469.

2. A. A. Golubov and A. V. Ustinov, IEEE Trans. Magn. **23** (1987) 781.

3. A. N. Vystavkin, Yu. F. Drachevsky, V. P. Kosheletz and I. L. Serpuchenko, Fiz. Nizk. Temp., **14** (1988) 646.

4. H. Akoh, S. Sakai, A. Yagi and H. Hayakawa, IEEE Trans. Magn., **21** (1985) 737.

5. Yu. S. Kivshar, A. N. Kosevich and O. A. Chubykalo, Phys. Lett. A, **129** (1988) 449.

6. Yu. S. Kivshar, A. M. Kosevich and O. A. Chubykalo, submitted to Phys. Rev. B (1988).

7. N. Flytzanis, St. Pnevmatikos and M. Remoissenet, J. Phys. C., **18** (1985) 4603.

8. Q. Li, St. Pnevmatikos, E. N. Economou and C. M. Soukoulis, Phys. Rev. B, **37** (1988) 3534.

Space Stochastic Perturbations of a Sine-Gordon Soliton

P.J. Pascual[1], *L. Vazquez*[1,3], *S. Pnevmatikos*[2,3], *and A.R. Bishop*[3]

[1]Departamento de Fisica Teorica, Facultad de Ciencias Fisicas,
 Universidad Complutense, E-28040 Madrid, Spain
[2]Research Center of Crete, P.O. Box 1527, GR-711 10 Heraklio,
 Crete, Greece
[3]CNLS, Los Alamos National Laboratory, Los Alamos, NM 87545, USA

1. INTRODUCTION

It is well known that (1+1) dimensional Sine-Gordon solitons represent structures with remarkable stability properties, and they can be used as models of ˝relativistic˝ extended particles. In this context the study of the dynamics of the Sine-Gordon solitons under stochastic perturbations is of great interest [1,2]. Such perturbations can simulate the interaction with a fluctuating external field or the modulation of the substrate potential barrier height as well as the effect of thermal noise or impurities.

More generally, the stochastic Sine-Gordon model can be considered from at least four important perspectives:

1. As a model incorporating balancing effects between nonlinearity and disorder; both can produce localization but on different (possibly competing) length scales.
2. As the brownian motion of an extended particle, which might be incorporated in a Fokker-Planck approach with appropriate collective coordinates.
3. Since the pure Sine-Gordon system is integrable, the associated stochastic system might be considered in terms of a stochastic KAM theorem for infinite degrees of freedom.
4. From the mathematical point of view, the model represents an excellent example of systematic study of stochastic nonlinear differential equations.

The present work is a short report of a preliminary study of such questions. The strategy of our approach is the following :

A. Numerical study. We numerically integrate the stochastic nonlinear wave equation using a finite difference scheme [1,3]. The time average of relevant quantities are evaluated over a finite set of ˝trajectories˝ associated with the response of the system to particular realizations of the stochastic perturbation.

B. Perturbative approach. For weak stochastic perturbations we assume that the dominant effect is to modulate the principal particle-like characteristics of the solitary waves. This allows us to isolate the main coordinates of the soliton and reduce the nonlinear partial differential equation (PDE) to a set of stochastic nonlinear ordinary differential equations (ODEs).

C. Mathematical study. Analysis of the convergence and stability of the stochastic numerical schemes. It is important to know that the scheme when applied to the deterministic part of the equation is not chaotic. Otherwise, we have some difficulties to distinguish the external from internal numerical irregularity and isolate the effect of noise.

Springer Proceedings in Physics, Vol. 39   **Disorder and Nonlinearity**
Editor: A.R. Bishop   © Springer-Verlag Berlin, Heidelberg 1989

2. COLLECTIVE COORDINATE ANALYSIS

The stochastic system we are dealing here is the perturbed Sine–Gordon equation:

$$\varphi_{tt} - \varphi_{xx} + [1+V(x)] \sin\varphi + F(x) = 0 \tag{1}$$

where $V(x)$ and $F(x)$ are referred to as additive and multiplicative noise, respectively, and both of them are static Gaussian white noises, localized in space and with vanishing average, i.e.

$$V(x), \quad F(x) \quad = \quad S(x) \quad , \quad x \in [-L/2, L/2]$$
$$= \quad 0 \quad , \quad x \notin [-L/2, L/2] \tag{2a}$$

with $\quad <S(x)> = 0 \quad$ and $\quad <S(x)\, S(y)> = 2D\delta(x-y).$ \hspace{1em} (2b)

The stationary stochastic solutions of (1) had been studied elsewhere [4]. Here, we report the behaviour of a single kink soliton, initially at rest at the origin, and surrounded by a stationary impurity region (spatial disorder) with an extent which is very large compared to the width of the soliton. For $F(x)=0$, the spatial disorder appears in equation (1) only multiplicatively and modulates stochastically the coefficient of the nonlinear term. This physically can be due to the inhomogeneous nature of the isolant in a long Josephson junction sandwich [5] or to a random modulation of the substrate potential in a Frenkel– Kontorova atomic chain, which in the continuum limit is also described by equation (1).

In Figure 1, we represent a panoramic spatiotemporal evolution of the Sine-Gordon kink under spatial white noise. The integration is done on the partial differential equation (1) and σ is defined by the relation : $D = \sigma^2 \Delta x/2$.

Linear perturbation analysis, for $F=0$, gives the dynamical equations [1,3] :

$$\frac{dX}{dt} = U + \frac{1}{4} U (1-U^2)^{1/2} \int_{-\infty}^{\infty} V(x)\,\theta\,\sin\varphi(\theta)\,\mathrm{sech}\theta\, dx \tag{3a}$$

$$\frac{dU}{dt} = \frac{1}{4} (1-U^2) \int_{-\infty}^{\infty} V(x)\,\sin\varphi(\theta)\,\mathrm{sech}\theta\, dx \tag{3b}$$

$$\theta = x - X / (1-U^2)^{1/2} \tag{3c}$$

where X is the center and U the velocity of the kink. Equations (3) represent a particle motion in a random force field depending on the position, which automatically corresponds to a problem of stochastic acceleration [6]. We can analyze some features of (3) by using a perturbative approach in the non-relativistic limit ($U^2 \ll 1$) yielding :

$$\frac{dX}{dt} = U + U\, H(X) \tag{4a}$$

$$\frac{dU}{dt} = G(X) \tag{4b}$$

where

$$H(X) = \int_{-\infty}^{\infty} V(x)\, h(x-X)\, dx \tag{5a}$$

$$G(X) = \int_{-\infty}^{\infty} V(x)\, g(x-X)dx \tag{5b}$$

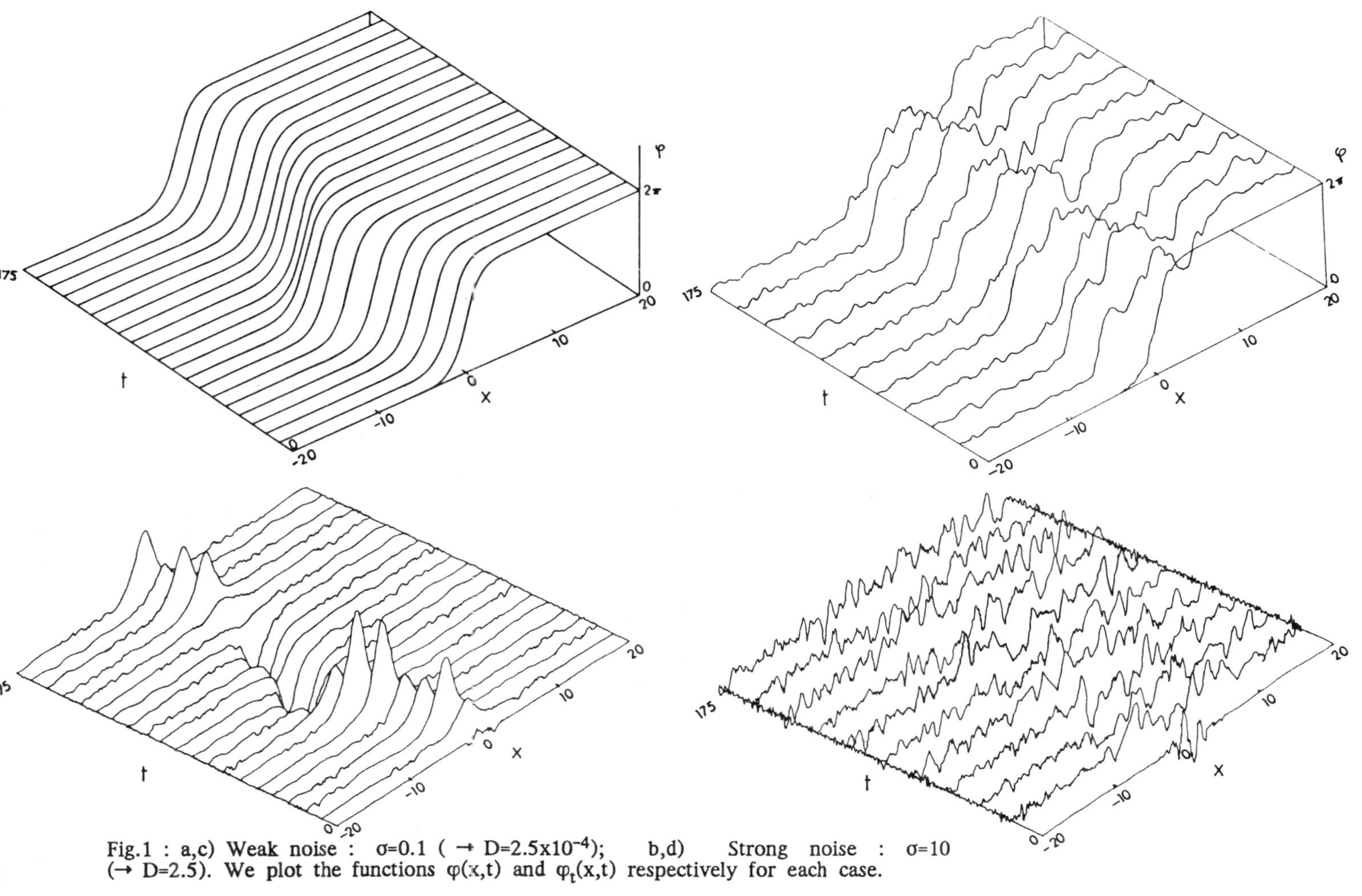

Fig.1 : a,c) Weak noise : $\sigma=0.1$ ($\rightarrow$ D=2.5x10^{-4}); b,d) Strong noise : $\sigma=10$ ($\rightarrow$ D=2.5). We plot the functions $\varphi(x,t)$ and $\varphi_t(x,t)$ respectively for each case.

are stochastic integrals, and the functions h and g are :

$$g(x) = \frac{1}{4} \sin\varphi(x) \ \mathrm{sech}(x) \tag{6a}$$

$$h(x) = xg(x) \tag{6b}$$

$$\varphi(x) = 4\tan^{-1} e^x \tag{6c}$$

with (6c) the Sine-Gordon kink soliton.

3. RESULTS

The set of equations (4) can easily be treated for three particular cases :

(a) The unperturbed case. This corresponds to the pure Sine-Gordon free soliton motion, where :

$$U{=}U_0 \quad \text{and} \quad X{=}X_0{+}U_0 t \tag{7}$$

is a solution of (4) in terms of mean values. The reason is that

$$<H(X_0{+}U_0 t)>{=}0 \quad , \quad <G(X_0{+}U_0 t)>{=}0 \tag{8}$$

according to (5) and the statistical properties of $V(x)$.

(b) Short time scale. In general, the system (4) can be written as follows :

$$d^2X/dt^2 = - \frac{dH(X)/dX}{1+H(X)} (dX/dt)^2 + G(X) \ [1+H(X)] \tag{9}$$

If we consider the soliton initially at rest at the origin ($X_0{=}0$), we can assume that, for short times, the dynamics is governed by the equation

$$d^2X/dt^2 = - \frac{dH(0)/dX}{1+H(0)} (dX/dt)^2 + G(0) \ [1+H(0)] \tag{10}$$

where $H(0)$, $G(0)$ and $dH(0)/dX$ are random variable, with statistical properties obtained from (2) and (5). The solution of (10) is :

$$X(t) = \frac{1}{\mu} \ \mathrm{Ln} \ \mathrm{ch} \ (\sqrt{\mu\lambda} \ t) \tag{11a}$$

$$U(t) = \frac{1}{\delta} \ (\frac{\lambda}{\mu})^{1/2} \ \mathrm{th} \ (\sqrt{\mu\lambda} \ t)) \tag{11b}$$

where μ, λ and δ are random variables

88

$$\mu = \frac{dH(0)/dX}{1+H(0)} \quad , \quad \lambda = G(0)\,[1+H(0)], \quad \delta = 1+H(0) \tag{12}$$

with the statistical properties :

$$<\mu>=0 \quad , \quad <\lambda>=0, \quad <\delta>=1 \tag{13}$$

$$<\mu\lambda> = 2D \int_{-\infty}^{\infty} g^2(x)dx>0 \rightarrow <\lambda/\mu>>0$$

Thus, we can verify that :

$$<U> \sim \alpha t$$
$$<X>_{FR} \sim \beta t^2 \qquad \text{as } U \rightarrow 0 \tag{14}$$

This statement is reflected on figures 2 and 3, and represents the asymmetric answer of the soliton to the stochastic perturbation.

(c) <u>Linear approximation</u>. The soliton is dynamically unstable (in a certain average) under the stochastic perturbation and in the framework of the collective coordinates. This can be seen by considering the linear approximation to the system (4) in the neighborhood of the solution (7), with $X_0=0$, $U_0=0$:

$$\frac{dX_I}{dt} = U_I + U_I\,H(0) \tag{15a}$$

$$\frac{dU_I}{dt} = G(0) - X_I dG(0)/dX \tag{15b}$$

The solution is :

$$X_I(t) = \frac{b}{d}\,[\,-\,ch\,\{t\,[-d(1+a)]^{1/2} + 1\}] \tag{16a}$$

$$U_I(t) = -\,b\,[-d(1+a)]^{-1/2}\,sh\,\{t\,[-d\,(1+a)]^{1/2}\} \tag{16b}$$

where a,b and d are the random variables

$$a=H(0) \quad , \quad b=G(0) \quad , \quad d=dG(0)/dX \tag{17}$$

with the statistical properties :

$$<a>=0 \quad , \quad <b>=0 \quad , \quad <d>=0$$
$$<-d(1+a)>= -<da> = D \int_{-\infty}^{\infty} g^2(x)dx>0 \tag{18}$$

According to (11) and (16) and with the values given by (13) and (18) we have two time scales with the relation:

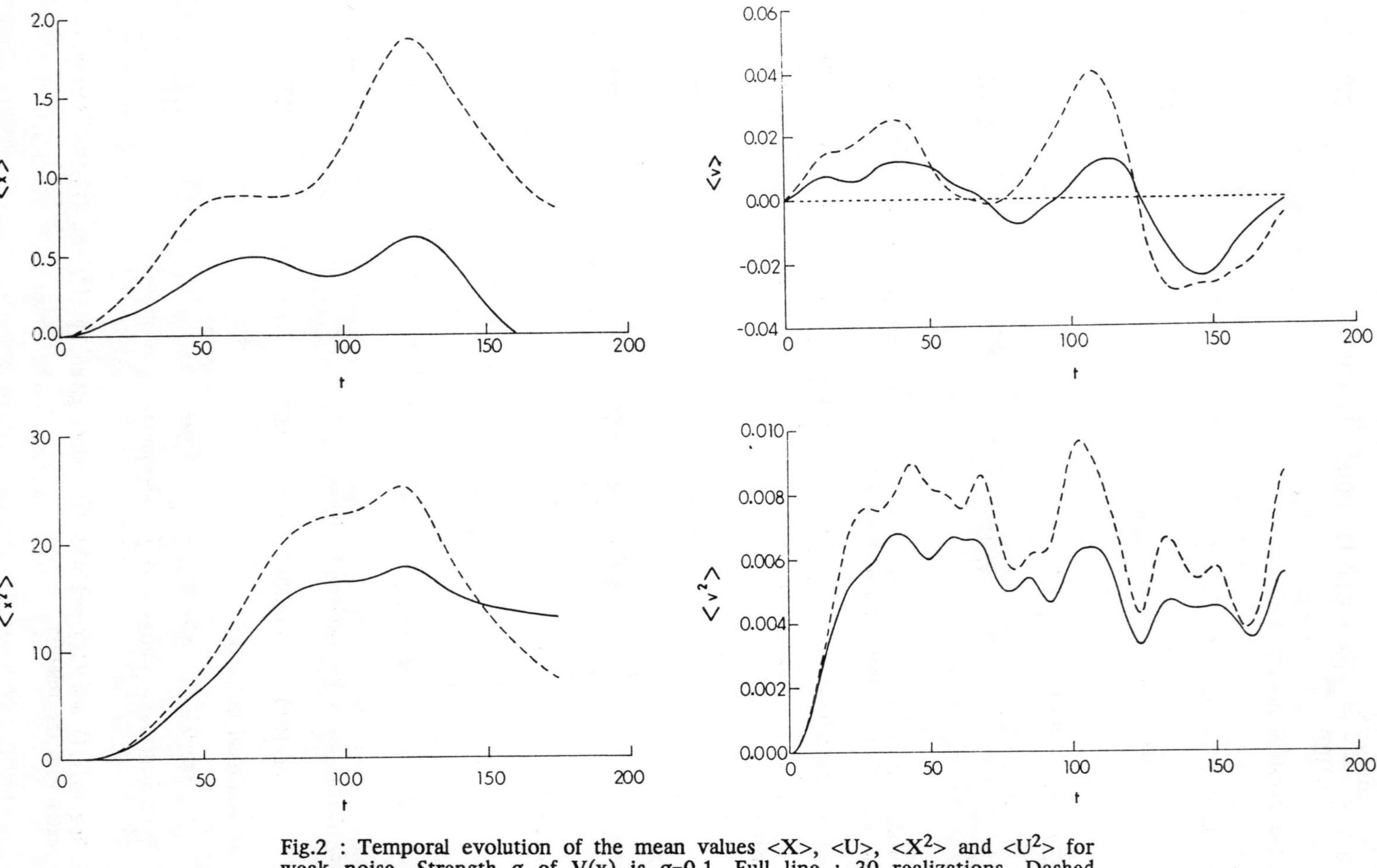

Fig.2 : Temporal evolution of the mean values $\langle X \rangle$, $\langle U \rangle$, $\langle X^2 \rangle$ and $\langle U^2 \rangle$ for weak noise. Strength σ of $V(x)$ is $\sigma=0.1$. Full line : 30 realizations. Dashed line : 15 realizations.

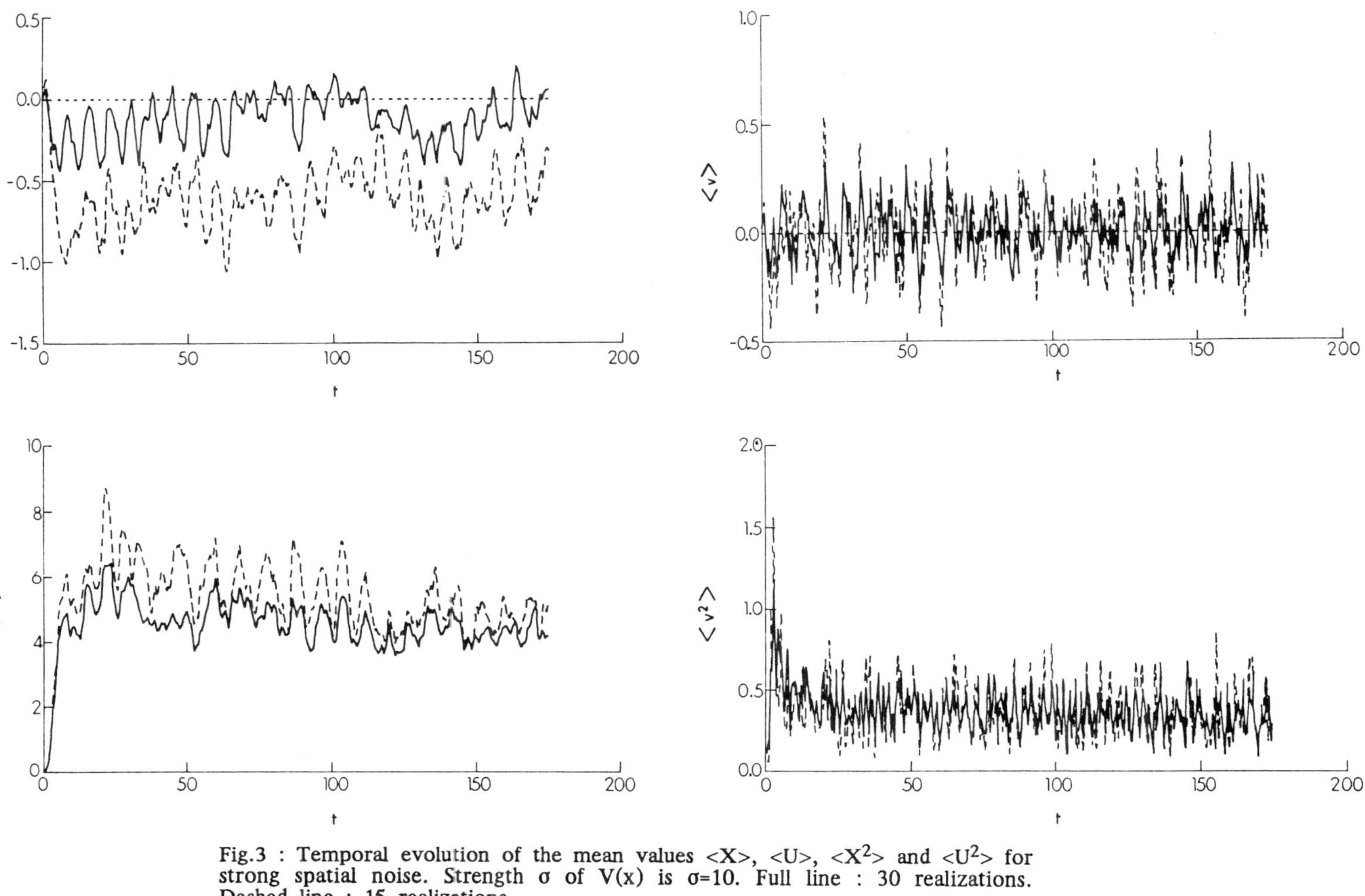

Fig.3 : Temporal evolution of the mean values $\langle X \rangle$, $\langle U \rangle$, $\langle X^2 \rangle$ and $\langle U^2 \rangle$ for strong spatial noise. Strength σ of $V(x)$ is $\sigma=10$. Full line : 30 realizations. Dashed line : 15 realizations.

$$[<1-da>]^{-1/2} / [<\mu\lambda>]^{-1/2} = 2^{1/2} \qquad (19)$$

On the other hand both scales behave like $D^{-1/2}$, where D is the strength of the noise.

In figures 2 and 3, we plot the mean values of $<X>$, $<U>$, $<X^2>$, $<U^2>$ for the case where the Sine-Gordon kink is initially at rest ($U_o=0$) and the spatial noise is included in the space interval $(-10, 10)$. These plots (numerical results) can be compared with our analytical formulas (11) and (16). Our results are obtained for both the weak : $D=2.5 \times 10^{-4}$ ($\sigma=0.1$) and strong : $D=2.5$ ($\sigma=10$) multiplicative white noise.

The present paper is only a short report of our preliminary results in the field of nonlinear partial differential equations with stochastic coefficients describing soliton properties in randomly perturbed physical media [7].

References

1. P.J. Pascual and L. Vazquez, Phys.Rev.B32, 8305 (1985).
2. F.G. Bass, Yu.S. Kivshar, V.V. Konotop and Yu.A. Sinitsyn, Phys.Rep.157, 63 (1988).
3. M.J. Rodriguez and L. Vazquez : "Stability of solitary waves under stochastic perturbations : A numerical and perturbative approach", in "Mathematics & Physics : Lectures on Recent Results", Vol.3, ed. L. Streit (World Scientific, in press).
4. A. Sanchez and L. Vazquez : "Stationary stochastic configurations of the Sine-Gordon system under spatial white noise", in preparation (1988).
5. G. Reinisch, J.C. Fernandez, N. Flytzanis, M. Taki and St. Pnevmatikos, Phys.Rev.B, in press (1988).
6. H. Kesten and G.C. Papanicolaou, Commun.Mathem.Phys.78, 19 (1980) and references therein.
7. P.J. Pascual, L. Vazquez, St. Pnevmatikos and A.R. Bishop, in preparation, (1988).

Chaotic Dynamics of Solitons and Breathers

F.K. Abdullaev and S.A. Darmanyan

Department of Thermophysics, Uzbek Academy of Sciences,
"Pravda" str. 28, SU-700135 Tashkent, USSR

1. Introduction

Nowadays the problem of chaotic dynamics of solitons and breathers in integrable and almost integrable systems [1–3] is of great interest. There are three main cases when we are investigating the chaotic dynamics of solitons.

We come across the first case when the chaotic dynamics arises due to the influence of random perturbations upon a system. Random perturbations always exist in nature and are caused by either random external fields or random inhomogeneities and nonstationary media. The soliton interaction with a thermal bath represents a typical example [4, 5].

A number of quantitative results can be obtained when the perturbation influence is small. In this case, we can develop the stochastic perturbation theory based on application of inverse scattering transform [6] or direct methods [7] and find the distribution function for soliton and breather parameters [1,2]. The stochastic perturbation theory also allowed us to describe the processes of wave radiation by solitons under the action of random perturbations and the appearance of radiative friction for solitons, etc.

The second case deals with the investigation of random field evolution in integrable and almost integrable systems, i.e. the evolution of random initial conditions. The first such treatment of the problem on the basis of an inverse scattering transform was made by Elgin and Kaup in application to the evolution of partial pulse radiation in an optical waveguide (it corresponds to the investigation of NSE).

Below, we consider the analogous problem for the KdV equation [7]. The SG and NSE have been studied in [8]. The problem of random field propagation in nonlinear dispersive media in the presence of amplification and random medium inhomogeneities requires further investigations.

The third case is connected with the appearance of soliton and breather chaotic dynamics when there are random peturbations neither in the initial conditions nor in the medium parameters. The corresponding equation represents regularly perturbed nonlinear differential equations in particular derivatives. The chaoticity in these systems arises due to the special types of instability of solitons under the action of periodic perturbations of arbitrary types of dissipation [9–12]. Here the inverse scattering transform is rather efficient. Below we consider a number of applications of this technique to the investigation of chaotization of soliton and breather motion of the perturbed SG equation. We can make use of the following procedure. First, with

Springer Proceedings in Physics, Vol. 39   **Disorder and Nonlinearity**
Editor: A.R. Bishop   © Springer-Verlag Berlin, Heidelberg 1989

the help of either asymptotic methods (the type of the perturbation theory based on the inverse scattering transform [13,14]) or direct methods, we find the system of ordinary coupled nonlinear differential equations for the soliton parameters. This system is finite dimensional. In the second stage, the methods of analysis of chaotic dynamics of finite-dimensional systems [16,17] are applied to this system. The inverse scattering transform scheme allows one to take into account the influence of soliton field radiation upon the conditions and mechanisms of the appearance of stochasticity [18].

Secondly, let us consider the problem of noise signal propagation in nonlinear media. The interest in the problem is due to the applications (e.g., the problem of soliton excitation by the partially coherent laser field in optical waveguides, etc.) as well as the general theory of nonlinear waves. There are a number of approximate approaches to the description of these problems. New possibilities for the description of random wave evolution in nonlinear dispersive media are connected with the inverse scattering transform [6]. On this basis, the noise signal transformation into solitons can be analytically described. Moreover, this method is rather efficient for the study of the evolution of random nonlinear dispersive wave packets in the presolitonic sector and problems of random signal propagation in randomly inhomogeneous and nonlinear active dispersive media (in the case of almost integrable systems). On the basis of the inverse scattering transform, the problems of intense electromagnetic pulse propagation with stochastic phase variation in nonlinear optical waveguides [3,19] are reduced to the investigation of the evolution of random initial conditions in the nonlinear Schrödinger equation and have recently been solved.

Let us consider the KdV equation giving the universal description of long wave dynamics in weakly nonlinear and weakly dispersive media [7]:

$$u_t - 6uu_x + u_{xxx} = 0 \quad , \tag{1}$$

with the initial conditions

$$u(x, t)|_{t=0} = u_s(x)[1 + \varepsilon(x)] \quad , \tag{2}$$

where

$$u_s(x) = 2q_0^2 \operatorname{sech}^2 q_0 x \quad ,$$

u_s is the single soliton solution, $\varepsilon \ll 1$ is the random gaussian function, and

$$\langle \varepsilon \rangle = 0, \quad \langle \varepsilon(x)\varepsilon(y) \rangle = B(x - y, l) \Rightarrow \sigma^2 \delta(x - y), \quad l \to 0 \quad ;$$

l is the correlation length.

According to the scheme of the inverse scattering transform it is necessary to find the characteristics of the spectrum of the linear operator in order to define the corrections to the scattering data under a random perturbation of the potential due to the correction to the potential $-2\varepsilon(x)q_0^2 \operatorname{sech}^2 q_0 x$

$$\Delta = \Delta_1 + \Delta_2 = q_0^3 \int_{-\infty}^{\infty} \frac{\varepsilon(x)dx}{\cosh^4 q_0 x}$$

$$-\frac{q_0^5}{\pi}\int\!\!\!\int\!\!\!\int\limits_{-\infty}^{\infty} dk\,dx\,dy\,\frac{\varepsilon(x)\varepsilon(y)\chi_k(x)\chi_k^*(y)}{(k^2+q^2)\cosh^3 q_0 x\,\cosh^3 q_0 y}\quad,\tag{3}$$

where

$$\chi = e^{-ikx}\,\frac{(ik+q_0\tanh q_0 x)}{ik-q_0}$$

is the Jost function of the Schrödinger equation with a potential $u_s(x)$. In the case of the δ-correlated random function $\varepsilon(x)$ we have

$$\langle \Delta \rangle = \langle \Delta_2 \rangle \approx -0.15\sigma^2 q_0^3$$
$$\langle \Delta \rangle = \langle \Delta_1^2 \rangle \approx 0.9\sigma^2 q_0^5\quad.\tag{4}$$

As shown by the analysis, in the soliton center the amplitude decreases during propagation in a medium, and it increases at the edges.

Next let us analyse the continuum components arising during the soliton formation from noise signals. We start from the calculation of $|R|^2$, where R is the reflection coefficient. Using the representation for R from [20]

$$R(k) = \frac{A(k,k)}{2ik-A(-k,k)}$$
$$A(q,k) = \int\limits_{-\infty}^{\infty} dx\,u(x)\chi_k(x)\,e^{-iqx}\tag{5}$$

we find at $k \leq q_0\alpha$, $\alpha^2 = \langle J_1^2 \rangle$,

$$J_1 = q_0\int\limits_{-\infty}^{\infty} dx\,\frac{\varepsilon(x)\tanh q_0 x}{\cosh^2 q_0 x}$$

that

$$|R|^2 = q_0^2 J_1^2(k^2+q_0^2 J_1^2)\quad.\tag{6}$$

At small R, $\ln(1-|R|^2) \approx -|R|^2$. For $B(x-y,l) = B_0\exp[-(x-y)^2/l^2]$, we obtain

$$\langle |R|^2 \rangle \approx \frac{q_0^2\alpha^2(5k^2+q_0^2)}{k^4+q_0^2 k^2+q_0^4\alpha^2}\,e^{-k^2 l^2}\quad.\tag{7}$$

Using (7) it is possible to calculate the continuum part of the integral invariants of the KdV equation, e.g. the total Hamiltonian. In this case we can show that the spectral density of energy has its maximum value at $k_c \sim l^{-1}$, i.e. being most efficiently excited in soliton tails of the waves with $k \sim k_c$.

As we have already noted, the chaotic dynamics of solitons and breathers can arise under the action of periodic peturbations of a general type caused by the influence of external fields, medium inhomogeneities, etc.

With the help of direct methods for solitons we can study the soliton motion in the perturbed u^4-model

$$u_{tt} - u_{xx} - u + u^3 + \Gamma u_t = f_0 \cos \omega t \quad . \tag{8}$$

It is necessary to perform the renormalization of the soliton field, using the representation

$$u = \Phi(x, t) + \varphi(t), \quad \varphi \ll \Phi \quad .$$

Substituting this expression into (8) one obtains the equation system for slowly changing field Φ and induced field $\varphi(t)$

$$\Phi_{tt} - \Phi_{xx} - \Phi + \Phi^3 + 3\Phi^2 \varphi = 0 \tag{9}$$

$$\varphi_{tt} - \varphi + \varphi^3 + \Gamma \varphi_t = f_0 \cos \omega t \quad . \tag{10}$$

Here we consider the damping to be small enough, i.e. $3\Phi^2 \varphi \gg \Gamma \varphi t$. From (10) we define the induced field $\varphi(t)$ and substituting it into (9) we find the closed equation for Φ. The field Φ takes the form

$$\Phi(x, t) = \tanh\left[\left(x - \int_0^t v(t')dt'\right) \bigg/ \sqrt{1 - v^2(t)}\right] \quad .$$

Applying perturbation theory we find the equation for the soliton velocity

$$\frac{dv}{dt} = -\frac{3\varphi(t)}{<\Phi_z^2>} \int_{-\infty}^{\infty} dz \Phi^2(z)\text{sech}^2 z \quad . \tag{11}$$

Using the results of [21] we find the Fourier transform of the correlation function for the soliton velocity

$$E_v(\omega) = \frac{g\pi^2}{2T\omega^2}\text{sech}^2(\pi^2\omega) \tag{12}$$

where T is given by

$$T \approx \left(1 + \frac{\Gamma}{2}\right) \ln\left\{ f_0\omega \,\text{sech}\left(\tfrac{1}{2}\pi\omega\right)\right\} \quad . \tag{13}$$

Next we study the solitonic analogue of the Ulam problem for the SG soliton. The corresponding problem represents the dynamics of the SG soliton motion between two delta-potential barriers. One of these barriers is periodically oscillating in space. The wave equation describing the system takes the form

$$\varphi_{tt} - \varphi_{xx} + \sin \varphi = -\varepsilon[\delta(x - L) + \delta(x + L + a \sin \omega t)]\sin \varphi \quad . \tag{14}$$

The evolution of soliton parameters is described by the system of equations for the soliton velocity $v(t)$ and the coordinate of the soliton center $\xi(t)$

$$\frac{dv}{dt} = \frac{\varepsilon}{2}\left[(1 - v^2)\mathrm{sech}^2\left(\frac{\xi - L}{\sqrt{1 - v^2}}\right)\tanh\left(\frac{\xi - L}{\sqrt{1 - v^2}}\right)\right.$$
$$\left. + (1 - v^2)\mathrm{sech}^2\left(\frac{\xi + L + a\sin\omega t}{\sqrt{1 - v^2}}\right)\tanh\left(\frac{\xi + L + a\sin\omega t}{\sqrt{1 - v^2}}\right)\right] \tag{15}$$

$$\frac{d\phi}{dt} = v - \frac{\varepsilon}{2}\xi v\left[\mathrm{sech}^2\left(\frac{\xi + L + a\sin\omega t}{\sqrt{1 - v^2}}\right)\tanh\left(\frac{\xi + L + a\sin\omega t}{\sqrt{1 - v^2}}\right)\right] \ . \tag{16}$$

For simplicity we study the case of small soliton velocity. The equation for the soliton center then reads

$$\frac{d\xi}{dt} = \pm\sqrt{H - 2\varepsilon\frac{(\alpha y + 1)}{(y + \alpha)^2}} \ , \tag{17}$$

where $\cosh^2\xi = y$, $\cosh^2 L = \alpha$.

We also investigate the solitonic behavior near the separatrix. The separatrix is defined by the condition

$$y_0 = \alpha - 2/\alpha, \quad H_c = \frac{\varepsilon\alpha^2}{2(\alpha^2 - 1)} \ , \tag{18}$$

where H_c is the value of the total hamiltonian on the separatrix. From (17) we find

$$\sqrt{H_c}(t - t_0) = -x_0 + \frac{\alpha + y_0}{2\sqrt{y_0^2 - 1}}\cdot\ln\left[\frac{y_0 - 1 + \sqrt{y_0^2 - 1}\tanh x}{y_0 - 1 - \sqrt{y_0^2 - 1}\tanh x}\right] \ . \tag{19}$$

To investigate the chaotic dynamics of the soliton it is convenient to use the Melnikov method [22]. Let us calculate the Melnikov function $M(t_0)$ characterizing the stochastic layer width near the separatrix. The calculation gives us the result

$$M(t_0) \simeq \cos(\omega t_0 + \omega L + \varphi)\frac{\pi\varepsilon a\omega}{\sinh\pi\omega/2} \ . \tag{20}$$

We then find that the maximal value occurs when $\omega_c \simeq 0.6$. If the initial data satisfy (20) then the soliton motion is chaotic and the soliton may be stochastically accelerated and may overcome the potential barrier. This is the difference from the Fermi-Ulam problem where the barrier is inpenetrable.

In conclusion we consider the dynamics of the rotationally symmetric solution of the sine-Gordon equation in a variable parametrically acting field. The SG equation in the rotationally symmetric case takes the form

$$\varphi_{tt} - \varphi_{zz} + \sin\varphi = \frac{D}{r}\varphi_2 - \varepsilon\sin(\omega t)\sin\varphi \ . \tag{21}$$

The calculation showed that the kink solutions represent the oscillating bubbles that after several oscillations (usually 5-10) are transformed into the emission field. Here we study the initial stage of the kink evolution in a high frequency field in the adiabatic approximation [23].

In general we apply the approach developed in the paper of M.E. Maslov [24]. To analyse the next stages of kink evolution it is necessary to take into account the interaction of the variable field with the kink emission field. It requires a separate investigation. Let us consider the kink solution of (21) when $\varepsilon = 0$ and $D = 2$ (three-dimensional case). It takes the form

$$\varphi_s = 4\tan^{-1}\left[\exp\left(\frac{r-p}{\sqrt{1-v^2}}\right)\right] \quad .$$
(22)

At $\varepsilon \neq 0$ the system of equation for the kink parameters takes the form

$$\frac{dv}{dt} = -\frac{2}{\varrho}(1-v^2)$$
(23)

$$\frac{d\varrho}{dt} = v + \frac{\varepsilon}{8}v(1-v^2)^{1/2}\sin\omega t$$

$$\varrho(t) = \varrho_0 \mathrm{cn}(\sqrt{2}t/\varrho_0, 1/\sqrt{2}) \quad (\text{at } \varepsilon = 0) \quad ,$$
(24)

where $\mathrm{cn}(x)$ is the Jacoby elliptic function. This system can be written in Hamiltonian form ($\varepsilon = 0$) with the free Hamiltonian

$$H_0 = \ln(\varrho^2 \cosh p), \quad p = \tanh^{-1} v \quad .$$
(25)

It is convenient to consider action-angle variables (I, θ). We obtain

$$I(H_0) \simeq A\exp(H_0)/2$$

$$A = \frac{K(1/\sqrt{2})}{\sqrt{2}\pi} \quad , \quad H_0 = 2\ln\frac{I}{A} \quad ,$$

$$\omega(I) = \frac{\partial H_0}{\partial I} = \frac{2}{I} \quad .$$
(26)

The condition causing the nonlinear resonances may be found by standard methods (see [16])

$$\frac{m\omega(I)}{2} = x \quad .$$
(27)

The distance between resonances $\delta\omega$ is

$$\delta\omega \simeq \frac{2}{x I_0^2} \quad .$$
(28)

The region of localization of the nonlinear resonance $\Delta\omega$ is equal to

$$\Delta\omega \approx \frac{4}{K^2(1/\sqrt{2})}\sqrt{\frac{\varepsilon x}{I_0^3}}\, l^{-\alpha x/2I_0} \quad .$$
(29)

Applying the Chirikov criterion [16] we find from (27-29) the condition for dynamical stochastization of the kink motion

$$I_0^4 \sim \frac{K^4(1/\sqrt{2})\exp(\alpha\beta)}{4\varepsilon\beta^3} \quad , \quad \beta = \frac{x}{I_0} \quad . \tag{30}$$

We report the results of chaos dynamics from solitons and breathers. We have considered the SG soliton in periodically inhomogeneous media under a periodic external force in [26]:

$$\varphi_{tt} - \varphi_{xx} + (1 + \varepsilon\cos ux)\sin \omega t = f \sin \Omega t - \Gamma\varphi_t \quad . \tag{31}$$

This problem may be analysed by the Melnikov method. The Melnikov function $M(t_0)$ is equal to

$$M(t_0) = 8\left(\frac{a}{u^3}\right)^{1/2} + \frac{\pi^2 f \sin \Omega t_0}{2u\cosh \pi\Omega/2\sqrt{au}}$$

$$a = \frac{\pi\varepsilon u^2}{\sinh \pi u/2} \quad . \tag{32}$$

The soliton velocity then becomes a random function of time if

$$\frac{16\sqrt{au}\,\Gamma}{\pi^2 f}\cosh\frac{\pi\Omega}{2\sqrt{au}} < 1 \quad . \tag{33}$$

We also may find the spectral density of waves emitted by the soliton in the chaotic region [26]. Also there exists stochastic dynamics of the SG breather under a parametrically acting periodic force [27]. For the one-dimensional case (21) we obtain the criterion of stochasticity

$$K = \frac{4\varepsilon_0\Omega^2}{\cos^3 \gamma} > 1$$

where $\gamma = \tan^{-1}(\eta/\nu)$. The boundary of the stochastic layer is given by

$$\overline{\gamma} = \arccos(4\varepsilon_0\Omega^2)^{1/3} \tag{35}$$

and its width near the separatrix $\gamma_s = \pi/2$ is

$$|\gamma_s - \overline{\gamma}| = (4\varepsilon_0\Omega^2)^{1/3} \quad . \tag{36}$$

If the initial breather data satisfy the condition (36), then the breather under the action of a periodically parametric perturbation oscillates randomly and its dynamics becomes diffusive. As a result, the breather decays into a free kink-antikink pair.

It is necessary to note that there is an analogy between linear quantum stochastic problems and nonlinear classical wave problems. Formally it is the analogy between linear stochastic differential equations in particular derivatives and regular nonlinear differential equations in particular derivatives. Both approaches lead to space localization of the solutions – Anderson localization and solitons. In 1979 one of the authors of this paper (F.Kh. Abdullaev together with S.S. Abdullaev) noted that such a phenomenon as Anderson localization has a wave character and can be observed in the classical physics of waves as well as in the quantum region [28]. This work considered beam propagation in the system of tunnel coupled optical waveguides: plane waveguides and fibers.

The equation for the mode parameters takes the form

$$-i\frac{da_n^i}{dz} = K_n^i a_n^i + \sum_{j,n \neq n'} V_{nn'}^{ij} a_{n'}^j \quad ,$$

where K_n^i is the propagation constant for the ith mode of the nth fiber, $V_{nn'}^{ij}$ is the coupling constant of i and j modes in the n and n' fibers. We assume that K_n are random quantities with the distribution function $P(K_n)$ having width W.

This model coincides with the Anderson model, where the coordinate z plays the role of time, K_n are the energies E_n, and the transitions between different sites in the lattice are described by $N_{nn'}$. The calculation shows that there exists a threshold K_c at which localization of light in one optical waveguide occurs (the optical analogue of Anderson localization).

In conclusion we point out that in [29] the possibility of stochastic parametric soliton resonance has been demonstrated. It is known that oscillating systems can have parametric amplification of the fluctuations of the system parameters. The amplification is connected with the fact that the random process corresponding to the fluctuations contains all the harmonics, and some of them correspond to the condition of parametric resonance. In [29], on the basis of nonlinear systems described by the NSE and KdV equations with fluctuating parameters, it has been shown that the analogous phenomenon can be observed for solitons.

References

1. F.Kh. Abdullaev, P.K. Khabybullaev: *Dynamics of Solitons in Inhomogeneous Condensed Matter* (FAN, Tashkent 1986)
2. F. Bass et al.: Phys. Rep. **157**, 63 (1988)
3. D. Kaup: In *Dynamical Problems in Soliton Systems* (Springer, Berlin, Heidelberg 1985) p.12
4. F.Kh. Abdullaev et al.: Phys. Status Solidi (b) **120**, 33 (1933)
5. F.Kh. Abdullaev: Fiz. Metallov **57**, 450 (1984) (in Russian)
6. V.E. Zakharov et al.: *Soliton Theory, Inverse Scattering Transform* (Nauka, Moscow 1989) (in Russian)
7. F.Kh. Abdullaev, S.A. Darmanyan: Zh. Tekh. Fiz. **58**, 265 (1988) (in Russian); Proc. International Conference on Nonlinear Acoustics (Novosibirsk, 1987)
8. F.Kh. Abdullaev et al.: *Optical Solitons* (FAN, Tashkent 1987) (in Russian)
9. F.Kh. Abdullaev: Lebedev Institute Reports **1**, 3 (1983)
10. I. Aranson, K. Gorshkov, M. Raynovich: ZhETPh. **86**, 929 (1984)
11. K. Nozaki: Phys. Rev. Lett. **49**, 1883 (1982)
12. A.R. Bishop et al.: Phys. Rev. Lett. **50**, 1095 (1983)
13. V.I. Karpman, E.M. Maslov: Phys. Lett. **60A**, 307 (1977)
14. D.J. Kaup, A.C. Newell: Proc. R. Soc., London **A361**, 413 (1978)
15. K.A. Gorshkov, L.A. Ostrovsky: Solitons in nonintegrable systems, preprint IPF AN SSSR, N12 (1981)

16. B.V. Chirikov: Phys. Rep. **52**, 278 (1979)
17. A.J. Lichtenberg, M.A. Lieberman: *Regular and Stochastic Motion* (Springer, Berlin, Heidelberg 1983)
18. K.Nozaki, Bekki: J. Phys. Soc. Jap. **54**, 2363 (1985)
19. J.N. Elgin: Phys. Lett. **110A**, 441 (1985)
20. F. Calogero, De Gasperis: *Spectral Transform and Solitons* (Academic, New York 1985)
21. V. Bruhsden, P. Holmes: Phys. Rev. Lett. **58**, 1699 (1989)
22. V. Melnikov: Trans. Moscow Math. Soc. **12**, 1 (1963)
23. F.Kh. Abdullaev, A. Umarov: Izv. VUZov-Radyofizica (1989) in press (in Russian)
24. E.M. Maslov: Physica **15D** (1985) 433-443
25. M.R. Samuelson: Phys. Lett. **74A** (1979) 21
26. F.Kh. Abdullaev et al.: DAN UzSSR, **N12**, 21 (1988)
27. F.Kh. Abdullaev et al.: Phys. Lett. **108A**, N 1, 51 (1985)
28. F.Kh. Abdullaev, S.S. Abdullaev: Izv. VUZov-Radyifizica **23**, 766 (1980), preprint INP, Tashkent (1979)
29. F.Kh. Abdullaev, S.A. Darmanyan, R. Dgumaev: Izv. AN UzSSR, N6, 53 (1986)

Proton Solitons in Hydrogen-Bonded Networks

G.P. Tsironis[1], S. Pnevmatikos[2,3], and P.S. Lomdahl[2]

[1]Institute for Nonlinear Science and Department of Chemistry,
 B-040, University of California, San Diego, La Jolla, CA 92093, USA
[2]Center for Nonlinear Studies, Los Alamos National Laboratory,
 Los Alamos, NM 87545, USA
[3]Research Center of Crete, P.O. Box 1527, GR-71 110 Heraklio,
 Crete, Greece

A two component soliton model is proposed for the study of defect formation and dynamics in a hydrogen-bonded network. Both ionic and orientational defects are described as distinct topological excitations in a doubly periodic substrate potential. Numerical studies of the interaction properties of both types of solitons present in the model are discussed.

1. Introduction

The electrical properties of hydrogen-bonded networks, and ice in particular, have been studied extensively over the last fifty years. It has been demonstrated that conductivity in ice crystals is not electronic but protonic in nature and that the protons that bond the oxygen atoms together are responsible for it [1]. The dynamics of the bonding protons has been studied in several theoretical ways, most of these have been based on Onsager's hopping model [2]. Recently, novel mechanisms for proton dynamics were proposed that take into account the cooperative nature of the motion of the protons in the system. The potential for the binding proton is that of a double well; this fact is responsible for the creation of collective topological excitations that play the role of ionic defects known to be present in ice [3-6].

A typical quasi-one dimensional hydrogen-bonded network is formed when a covalently bonded dimer ($X-H$) of a proton (H) with an ion (X) couples to an adjacent dimer via a hydrogen bond ($X-H \cdots X-H$). The resulting chain has the form:

$$\cdots X-H \cdots X-H \cdots X-H \cdots X-H \cdots$$

The most common example of such a system is provided by crystaline ice Ih, where X is an oxygen atom and the hydrogen bond connects the water molecules among themselves. In this case, quasi-one dimensional chains can be identified, termed Bernal-Fowler filaments [7], along which the dc-conductivity is substantially larger than along the other directions in the crystal. The generally accepted mechanism for charge transfer along such a filament is as follows: In equilibrium, each proton is surrounded by two neighboring ions. There are two equivalent equilibrium positions for each such proton, one corresponding to the covalent bond and the other to the hydrogen bond of the proton with each ion- the proton potential thus being bistable. When the chain is in its ground state the protons can be either in the left minimum of the potential (left groung state) or in the right minimum (right ground state). Excitations in the system involve transitions from an all-left to an all-right state; this marks the creation of an <u>ionic defect</u> with an excessive positive charge around the ion close to which the proton moved (creation of H_3O^+ in ice) and more negative charge in the vicinity of the original location of the proton (creation of OH^- in ice). This displaced protonic charge can be transfered to a neighboring hydrogen bond through a rotation of the molecule. As a result, two <u>orientational defects</u> are formed that involve bonds with no (L-defect) or two (D-defect) protons.

Springer Proceedings in Physics, Vol. 39   **Disorder and Nonlinearity**
Editor: A.R. Bishop   © Springer-Verlag Berlin, Heidelberg 1989

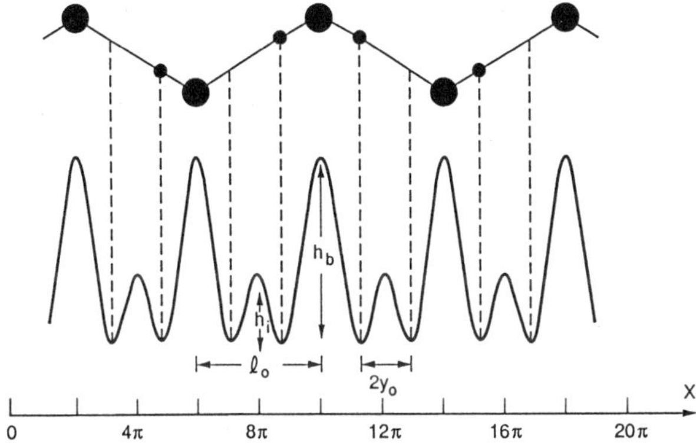

Figure 1. The dynamics of protons in a hydrogen-bonded chain is modeled with the help of a doubly periodic substrate potential.

A model has been introduced very recently in which both types of the aforementioned defects arise naturally as distinct topological solitons [5]. In Fig. 1 we plot the potential for the protons in a Bernal-Fowler filament. The larger barriers coincide with the equilibrium poisition of the ions whereas the minima separated by the smaller barriers represent the hydrogen bond that connects the adjacent ions. When a proton moves across a large barrier, an orientational defect is formed, whereas tunneling across a small barrier results in an ionic defect. The relative values of the barriers represent the ratio of the activation energies for ionic over the orientational defects and should be determined from experimental data. In the present model, this ratio depends on one free parameter. The choice we adopted here corresponds to the present understanding of the Ih ice crystal. This can be different in other hydrogen bonded systems [8]. The collective dynamics of the protons in this doubly periodic substrate potential coupled with the motion of the ions describe the ionic and orientational defects in terms of nonlinear topological excitations.

2. Collective Proton Dynamics

A realistic model for the collective proton dynamics in a hydrogen-bonded network must satisfy various physical requirements. It must describe accurately the nature of the hydrogen bond between adjacent ions and allow for spontaneous formation of both ionic and orientational defects. The Hamiltonian for such a one-dimensional model has three parts, viz.:

$$H = H_p + H_o + H_i \tag{1}$$

where H_p is the Hamiltonian for the proton sublattice, H_o for the ion sublattice, and H_i is the interaction term between the two. We have:

$$H_p = \sum_n \left[\frac{1}{2} m \left(\frac{dy_n}{dt}\right)^2 + \frac{1}{2} K_1 (y_{n+1}-y_n)^2 + S_p V_1\left(\frac{4\pi}{l_o} y_n\right) \right] \tag{2a}$$

$$H_o = \sum_n \left[\frac{1}{2} M \left(\frac{dY_n}{dt}\right)^2 + \frac{1}{2} K_2 (Y_{n+1}-Y_n)^2 + S_o V_2\left(\frac{Y_n}{l_o}\right) \right] \tag{2b}$$

$$H_i = \chi \sum_n \left[(Y_n-Y_{n-1})\Phi\left(\frac{4\pi}{l_o} y_n\right) \right] \tag{2c}$$

In eqs. (2), m, M denote the mass of the protons and ions, respectively, K_1, K_2 the corresponding (nearest-neighbor) spring constants, and χ is the coupling parameter between the two interacting sublattices. The displacement y_n of the n-th proton is measured from the central unstable position in the hydrogen bond, i.e., from the middle of the bond that links the ions, whereas, Y_n, the displacement of the n-th ion, is measured from its equilibrium position. The equilibrium distance between two heavy ions is taken to be l_o. With the introduction of the quantities:

$$u_n = \frac{4\pi}{l_o} y_n \,, \qquad w_n = \frac{Y_n}{l_o} \tag{3}$$

the potentials in (2) can be written as

$$V_1(u_n) = \frac{2}{1-a^2}\left[\cos(\frac{u_n}{2})-a\right]^2, \qquad 0<a<1 \tag{4a}$$

$$V_2(w_n) = \frac{1}{2}w_n^2 \tag{4b}$$

$$\Phi(u_n) = \cos(\frac{u_n}{2}) - \cos(\frac{u_o}{2}), \qquad u_o = 2arccos(a) \tag{4c}$$

The potential $V_1(u_n)$ is the on-site potential for the proton sublattice (Fig. 1) and is chosen to satisfy the physical requirements imposed by the hydrogen-bonded network. The potential function $\Phi(u_n)$ determines the interaction between the two sublattices and $V_2(u_n)$ provides the ions with a linear restoring force driving them towards their equilibrium position.

The equations of motion for the Hamiltonian of eq. (2.2) can be written in dimensionless form as follows:

$$\frac{d^2 u_n}{d\tau^2} = \omega_1^2(u_{n+1}-2u_n+u_{n-1}) - \Omega_1^2\frac{dV_1(u_n)}{du_n} - \chi_1(w_n-w_{n-1})\frac{d\Phi(u_n)}{du_n} \tag{5a}$$

$$\frac{d^2 w_n}{d\tau^2} = \omega_2^2(w_{n+1}-2w_n+w_{n-1}) - \Omega_2^2 w_n + \chi_2[\Phi(u_{n+1})-\Phi(u_n)] \tag{5b}$$

If we introduce the following constants: $\varepsilon_o = 1.986\times10^{-23}J$ for energy, $t_o = \omega_2^{-1} = (M/K_2)^{\frac{1}{2}}$ for time, and l_o for length, the energy in the Hamiltonian is determined in cm^{-1}. This system of units is introduced in order to facilitate the numerical computations and to enable the comparison of the results with experiments [5]. With these definitions and for the values chosen for ice [5] we have $\upsilon_o=1$ and $c_o=11$.

In the continuum limit eqs. (5) become

$$u_{\tau\tau} - c_o^2 u_{xx} + \Omega_1^2\frac{dV_1}{du} + \chi_1 w_x\frac{d\Phi}{du} = 0 \tag{6a}$$

$$w_{\tau\tau} - \upsilon_o^2 w_{xx} + \Omega_2^2 w - \chi_2\frac{d\Phi}{dx} = 0 \tag{6b}$$

where x, τ are the dimensionless space and time variables, c_o, υ_o denote the speed of sound in the proton and ion lattice respectively, χ_1, χ_2 are proportional to χ and Ω_1, Ω_2 are proportional to $S_p^{\frac{1}{2}}$, $S_o^{\frac{1}{2}}$ respectively.

In the special case where $\Omega_2 = 0$ eq. (6a) reduces to the double-sine Gordon equation (DSG) for the protonic sublattice:

$$(\upsilon^2-c_o^2)u_{\xi\xi}+\varepsilon[-\sin u+2a\sin(\frac{u}{2})]=0 \tag{7a}$$

while for the heavy sublattice we have:

$$w_\xi = \frac{\chi_2}{\upsilon^2-\upsilon_o^2}[\cos(\frac{u}{2})-a] \tag{7b}$$

with

$$\varepsilon = \frac{\Omega_1^2}{1-a^2} - \frac{\chi_1\chi_2}{4v_o^2(1-\dfrac{v^2}{v_o^2})} \qquad (7c)$$

where $\xi = x - v\tau$ and v is the traveling wave velocity. The parameter ε defines an effective coefficient for the new barrier height. We note that the sign of ε depends on both the nonlinear coupling parameter χ^2 and the travelling wave velocity v.

The solutions of the double Sine-Gordon equation for $\chi=0$ are well known [9,10]; they consist of two kind of topological solitons (and the corresponding anti-solitons) which have been termed "small" and "large" kinks, respectively. The former correspond to a transition over the small barrier of the potential in Fig. 1, whereas, the latter describe the transition over the larger barrier. In the context of our present model for the hydrogen bonded quasi-one dimensional chains, these two kinds of kinks are naturally associated with ionic and orientational defects, respectively. For small values of the coupling parameter χ, ε decreases (for $v=0$); this has the effect of increasing the width of the solitons, decreasing their energy, and thus making the excitations more stable. For finite velocities, DSG-type propagation is observed accompanied by gaps in propagation for certain velocity ranges [5].

When $\Omega_2 \neq 0$ the interpretation furnished by the DSG picture remains valid; there is however a change in the shape of the corresponding deformation in the ionic sublattice that accompanies the protonic deformation [5]. In Fig. 2, we show the two types of topological defects present in the system and their associated ionic deformations for $\Omega_2 \neq 0$.

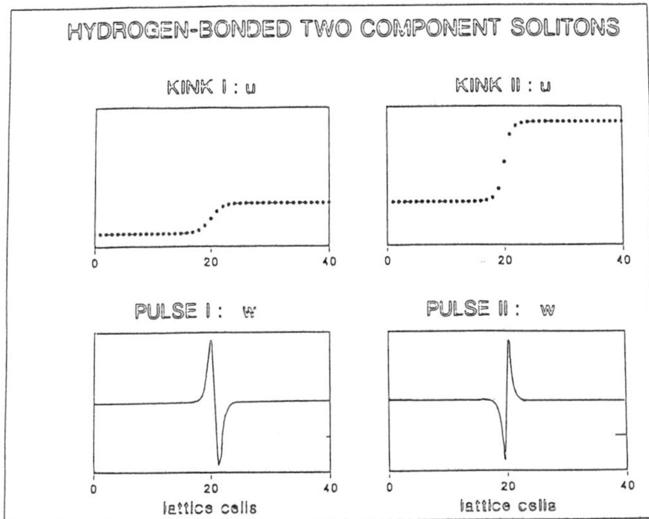

Figure 2. Small (ionic defects) and large (orientational defects) solitons and the associated deformation of the ions for $\Omega_2 \neq 0$.

3. Defect Propagation and Collision Properties

We recently performed a series of numerical experiments in order to analyze the collective dynamics of the nonlinear excitations of the model described in the previous section. The coupled set of equations in Eq. (5) was integrated for a one dimensional system containing 400 units. The exact analytical solutions of eq. (7a) were used as initial conditions for the computer simulations. Both the small and large kinks of the eq. (7a) were seen to propagate freely in the discrete system for most velocity ranges. The coupling of the protons with the ions was seen to hinder the kink motion as their velocity approached v_o, the speed of sound in the ion lattice. For kink velocities substantially larger than v_o, the effect of the interaction with the ionic lattice is substantially smaller.

The numerical study of the collision properties of the defects showed that the following types of reactions can occur: For velocities small compared to v_o, two colliding ionic defects (small kinks) mutually annihilate each other and leave in place a persisting oscillation. This localized oscillation, or breather, is formed because the kinks do not have enough energy to escape and therefore get trapped. For larger velocities, $(v/v_o)>1$, the two kinks collide almost elastically and reflect each other. This behavior is in agreement with the standard ϕ^4-soliton properties and arises from the fact that the ionic component of the substrate potential resembles a "truncated" ϕ^4-potential. The large kinks have more interesting behavior in the same large velocity range: when they collide they are seen to transform into pairs of small kinks that move in opposite directions. The analogous conversion of a small kink pair into a large one which been seen numerically in the pure DSG system was not observed here. In Figs. 3, 4 and 5 we present typical examples of collisions for small and large kinks.

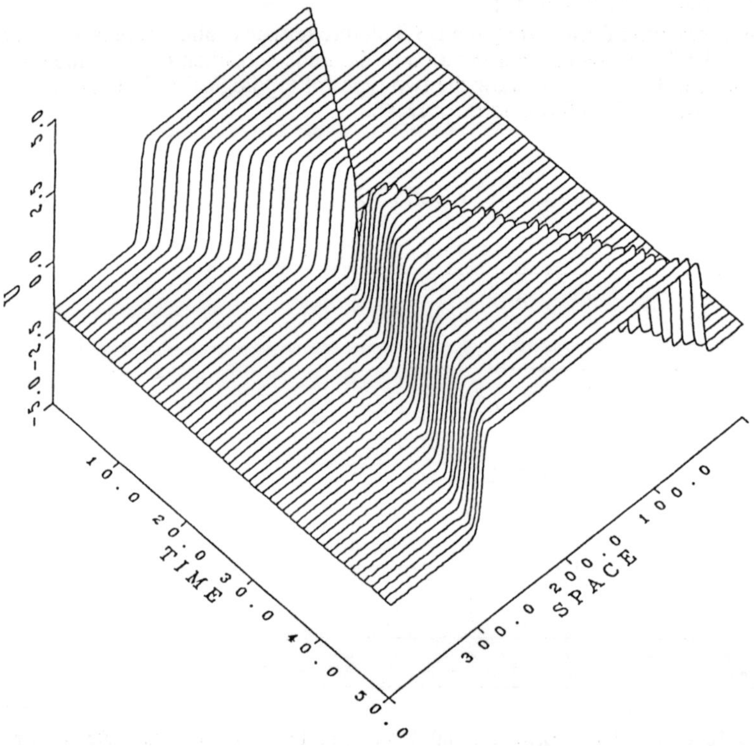

Figure 3. Elastic collision of two small kinks with initial velocities larger than v_o ($v_{100}=5.0\,v_o$, $v_{300}=-6.0\,v_o$; subscripts refer to initial positions of the solitons).

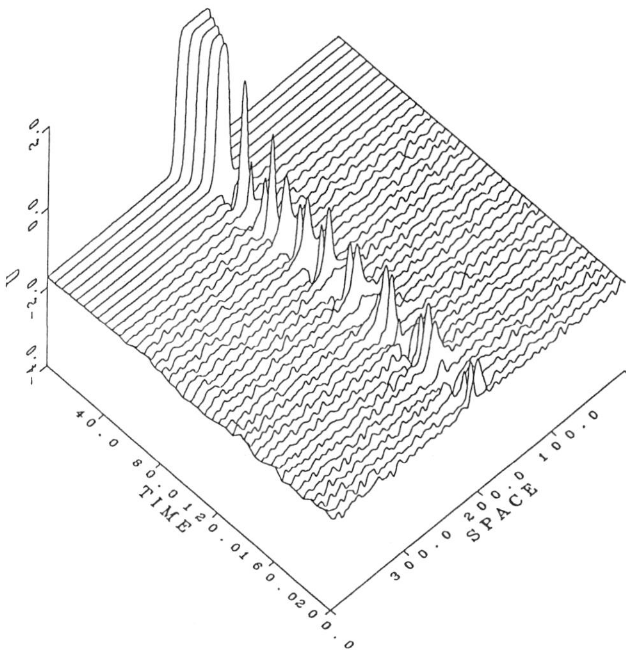

Figure 4. The collision of two small kinks with small initial velocities ($v_{175}=-v_{225}=0.8\,v_o$) results in a localized oscillatory mode.

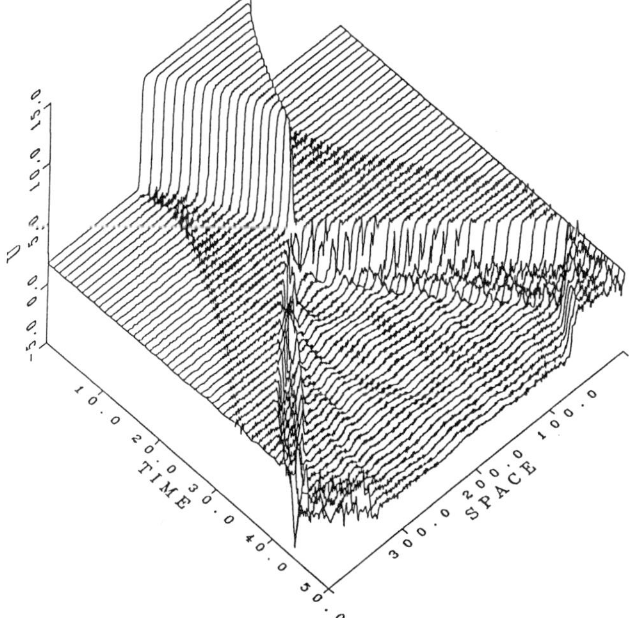

Figure 5. Head-on collision of two large kinks at high velocities. The mutual annihilation of the large kinks is accompanied by the creation of two small kinks traveling in opposite directions ($v_{100}=9.0\,v_o$, $v_{300}=-9.5\,v_o$).

Acknowledgements: We would like to thank Dr. Evita Vulgaris for help with the manuscript. One of us (G.P.T) acknowledges partial support of NSF grant DMR 86-19650-A1.

4. References

1.   P. B. Hobbs, Ice Physics, Clarendon Press, Oxford, 1974.

2.   L. Onsager, Science **156**, 541 (1967); **166**, 1359 (1969).

3.   J. H. Weiner and A. Askar, Nature, **226**, 842 (1970); V. Ya. Antonchenko, A. S. Davydov, and A. V. Zolotaryuk, Phys. Stat. Solidi (b) **115**, 631 (1983); E. W. Laedke, K. H. Spatschek, M. W. Wilkens Jr. and A. V. Zolotaryuk, Phys. Rev. A **32**, 1161 (1985); H. Weberpals and K. H. Spatscheck, Phys. Rev. A **36**, 2946 (1987); M. Peyrard, St. Pnevmatikos and N. Flytzanis, Phys. Rev. A **36**, 903 (1987).

4.   A. V. Zolotaryuk, K. H. Spatschek and E. W. Laedke, Phys. Let. A **101**, 517 (1984); St. Pnevmatikos, Phys. Lett. A **122**, 249 (1987); J. Halding and P. S. Lomdahl. Phys. Rev. A **37**, 2608 (1988).

5.   St. Pnevmatikos, Phys. Rev. Lett. **60**, 1534 (1988); G. P. Tsironis and St. Pnevmatikos, to be published, Phys. Rev. B; St. Pnevmatikos and G. P. Tsironis, in preparation.

6.   A. V. Zolotaryuk, Proceedings of the International Seminar on Biophysical Aspects of Cancer, ed. J. Fiala and J. Pokorny, Charles University, Prague 1987.

7.   J. D. Bernal and R. H. Fowler, J. Chem. Phys. **1**, 515 (1933).

8.   R. W. Jansen, R. Bertoncini, D. A. Pinnick, A. I. Katz, R. C. Hanson, O. F. Sankey and M. O' Keeffe, Phys. Rev. B **35**, 9830 (1987).

9.   K. M. DeLeonardis and S. E. Trullinger, Phys. Rev. B **27**, 1867 (1983).

10.   D. K. Campbell, M. Peyrard and P. Sodano, Physica D **19**, 165 (1986).

Part III

Scattering and Localization

The Statistics of Random Backscatter: A Comparison of Theory with Computer Simulations

G. Papanicolaou[1], M. Postel[1], P. Sheng[2], and B. White[2]

[1]Courant Institute of Mathematical Sciences, New York University,
 251 Mercer Street, New York, NY 10012, USA
[2]Exxon Research and Engineering Co., Route 22 East,
 Annandale, NJ 08801, USA

1 Introduction

We consider the statistics of the backscatter generated when an acoustic plane
wave pulse is normally incident on a randomly layered half space. The random
acoustic medium is assumed to have smooth long distance trends, with a small-scale
stochastic microstructure. The width of the incident pulse is chosen to be much
larger than the random microstructure but much smaller than the macroscopic variations. With this separation into three distinct spatial scales (microstructure $\ll$
pulse width $\ll$ macrostructure) a complete description of the backscatter statistics can be given for macroscopic times under wide hypotheses on the form of
the random variations.

The formulation and solution of the problem presented here follow [1]. The
backscatter is not stationary in time, but is approximately stationary over relatively short time segments. More specifically we consider a time window centered
at macroscopic time t, but of length comparable to the duration of the incident
pulse. Then within this window the backscatter is approximately a stationary
Gaussian process with a power spectrum that depends on t and certain averaged
properties of the random medium as explained further below. An earlier less explicit version of the theory is in [2], and related simulations and connections to
localization theory are in [3,4]. In this paper we present computer simulations
verifying some aspects of the theory. More extensive simulations and applications
to the stochastic inverse problem of determining the medium macroscopic variations
from backscattered data will be published elsewhere [5,6].

2 Formulation

We consider a one-dimensional acoustic wave propagating in a medium occupying $x<0$.
We will give statistics for the backscatter recorded at $x=0$.

Let $\rho(x)$ be density and $K(x)$ the bulk modulus. Then if $p(t,x)$ is pressure and
$u(t,x)$ velocity the acoustic equations are

$$\rho \frac{\partial u}{\partial t} + \frac{\partial p}{\partial x} = 0$$

$$\frac{1}{K} \frac{\partial p}{\partial t} + \frac{\partial u}{\partial x} = 0 \quad . \tag{1}$$

Let

$$\rho_0 = E[\rho]$$

$$K_0 = (E[1/K])^{-1} \tag{2}$$

In the special case that ρ, K are stationary random functions of position, ρ_0, K_0
are the constant parameters of effective medium theory. That is, a long wavelength
pulse will propagate over distances not too large as if in a homogeneous medium
with constant parameters ρ_0, K_0, and constant speed

Springer Proceedings in Physics, Vol. 39   **Disorder and Nonlinearity**
Editor: A.R. Bishop   © Springer-Verlag Berlin, Heidelberg 1989

$$c_0 = (K_0/\rho_0)^{1/2} \tag{3}$$

More generally ρ_0, K_0, c_0 are allowed to vary slowly compared to the spatial scale of the random microstructure, and are assumed to be differentiable functions of x.

Let ϵ, $0 < \epsilon \ll 1$ be a small parameter. Then if l_0 is the typical size of a macroscopic variation, we take $\epsilon^2 l_0$ to be the size of a typical random inhomogeneity. That is, ϵ^2 is the ratio of microscale to macroscale. This is expressed by putting

$$\rho(x) = \rho_0(x/l_0) \left[1 + \eta(x/l_0, \ x/\epsilon^2 l_0) \right]$$

$$\frac{1}{K(x)} = \frac{1}{K_0(x/l_0)} \left[1 + \nu(x/l_0, \ x/\epsilon^2 l_0) \right] \tag{4}$$

The random fluctuations η, ν have mean zero and slowly varying statistics.

For simplicity we consider here the case of a "matched medium". The random medium occupying $x<0$ is adjoined to a homogeneous medium for $x>0$ which has constant parameters $\rho_0(0)$, $K_0(0)$. The Green's function for this problem is then obtained from the initial conditions

$$u = l_0 \ \delta(t + x/c_0(0))$$

$$p = -l_0 \ \rho_0(0) \ c_0(0) \ \delta(t + x/c_0(0))$$
$$\text{for } t < 0 \quad . \tag{5}$$

The Green's function G will then be the right-going wave in $x>0$ as $x\downarrow 0$

$$G = -\frac{1}{2} \left[u(t,0) - \frac{p(t,0)}{(\rho_0(0) \ c_0(0))} \right] \quad . \tag{6}$$

Let $\bar{c}$ be a typical value of c. The incident pulse is taken to be of the form

$$f^\epsilon(t) = \frac{1}{\epsilon^{1/2}} \ f(\bar{c}t/\epsilon l_0) \quad . \tag{7}$$

f^ϵ therefore varies on the scale ϵl_0 which is large compared to the microscale $\epsilon^2 l_0$ but small compared to the macroscale l_0. The factor $\epsilon^{-1/2}$ is inserted to preserve energy as ϵ varies. By appropriate scaling, we may set $\bar{c}$, l_0 equal to unity.

The backscatter is now the convolution of the Green's function with the pulse. Let

$$G^\epsilon_{t,f}(\sigma) = (G * f^\epsilon) \ (t + \epsilon \ \sigma)$$
$$= \int_0^{t + \epsilon \ \sigma} G(t + \epsilon \ \sigma - s) \ f^\epsilon(s) \ ds \quad . \tag{8}$$

In analyzing the backscatter given by (8) we consider t, the "window center", to be a fixed macroscopic variable. The backscatter is considered a stochastic process in the "window variable" σ which varies on the scale of a pulse duration. The statistics of this process then vary with t, the location of the time segment being analyzed.

Let

$$\hat{f}(\omega) = \int_{\infty}^{\infty} e^{i\omega t} \ f(t) \ dt \tag{9}$$

be the Fourier transform of f and $\hat{G}$ an appropriately scaled Fourier transform lf G. Then from (8) we may write

$$G^{\epsilon}_{t,f}(\sigma) = \frac{1}{2\pi\epsilon^{1/2}} \int_{\infty}^{\infty} e^{-i\omega(t+\epsilon\sigma)/\epsilon} \hat{f}(\omega) \hat{G}(\omega) \, d\omega \quad . \tag{10}$$

Now (10) can be analyzed from knowledge of the statistics of $\hat{G}$. For $x<0$ let

$$\tau(x) = \int_{x}^{0} \frac{ds}{c_0(s)} \tag{11}$$

be the travel time to x in the macroscopic medium, and let

$$m = \frac{1}{2}(\eta + \nu)$$
$$n = \frac{1}{2}(\eta - \nu) \quad . \tag{12}$$

In [1], (10) is analyzed by deriving statistics for $\hat{G}$. It can be shown by an embedding argument that $G=\exp\{-\psi(0)\}$, where $\psi(x)$ is real and satisfies a nonlinear stochastic differential equation with small parameter ϵ. The results of the next section follow from asymptotic analysis of that nonlinear equation as $\epsilon\downarrow0$.

3 Backscatter statistics

In [1] it is shown that $G^{\epsilon}_{t,f}(\cdot)$ converges weakly as $\epsilon\downarrow0$ to a stationary, mean zero Gaussian process with power spectrum

$$S_t(\omega) = |\hat{f}(\omega)|^2 \, \mu(t,\omega) \quad . \tag{13}$$

The normalized spectrum μ is computed as follows. Let

$$\alpha_{nn}(x) = \int_0^{\infty} E\left[n(x,y) \, n(x,y+s) \right] ds \quad . \tag{14}$$

Let $\hat{x}(\tau)$ be the inverse of the travel time $\tau(x)$ given by (11). Let

$$\alpha(\tau) = \alpha_{nn}(\hat{x}(\tau))/c_0(\hat{x}(\tau)) \quad . \tag{15}$$

Let $W^N(\tau,t,\omega)$ be the solution of

$$\frac{\partial W^N}{\partial \tau} + 2N \frac{\partial W^N}{\partial t} - 2\omega^2 \alpha(\tau) \left\{ (N+1)^2 W^{N+1} - 2N^2 W^N \right.$$
$$\left. + (N-1)^2 W^{N-1} \right\} = 0$$
$$\text{for } t,\tau>0, \quad N=0,1,2,\ldots$$
$$W^N(0,t,\omega) = \delta(t) \, \delta_{N,1} \quad . \tag{16}$$

Then

$$\mu(t,\omega) = W^0(t/2,t,\omega) \quad . \tag{17}$$

In the special case that the medium is statistically homogeneous, i.e. ρ, K are stationary random functions, then α is constant and μ is given explicitly by

$$\mu(t,\omega) = \frac{\omega^2\alpha}{(1+\omega^2\alpha t)^2} \quad . \tag{18}$$

4 Comparison of Theory with Computer Simulations

Figures 1 and 2 illustrate the results of a stochastic simulation satisfying the separation of scales hypotheses, i.e., that the microscale $\ll$ pulse width $\ll$ macroscale. For these figures the random fluctuations η and ν were taken to be piece-wise constant over microlayers of 3 meters, and to be uniformly distributed in $[-.3,.3]$, with independent values in each layer. The incident pulse is a Ricker wavelet (second derivative of a Gaussian), and has a width of about 150 m, so that $\epsilon=.02$. The equations (1) were solved numerically using a finite difference scheme, adapted from [7], that is fourth order in space and second order in time.

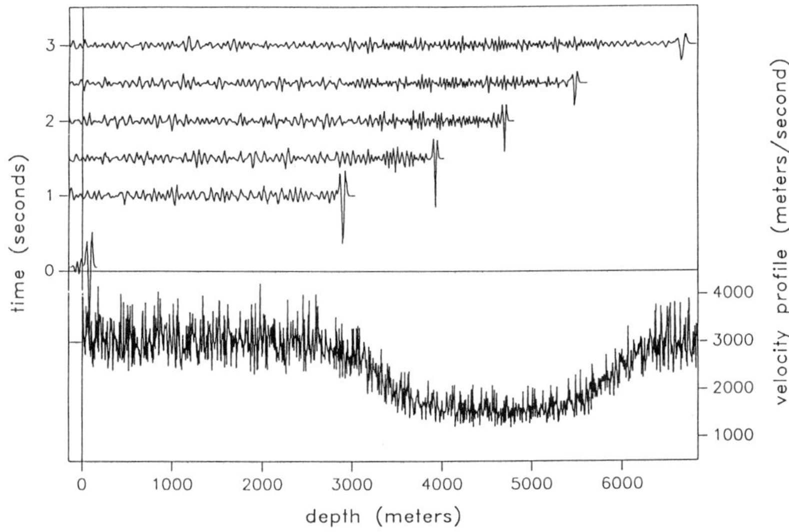

Figure 1. Velocity field and medium propagation speed. ϵ = .02, maximum noise= 30%

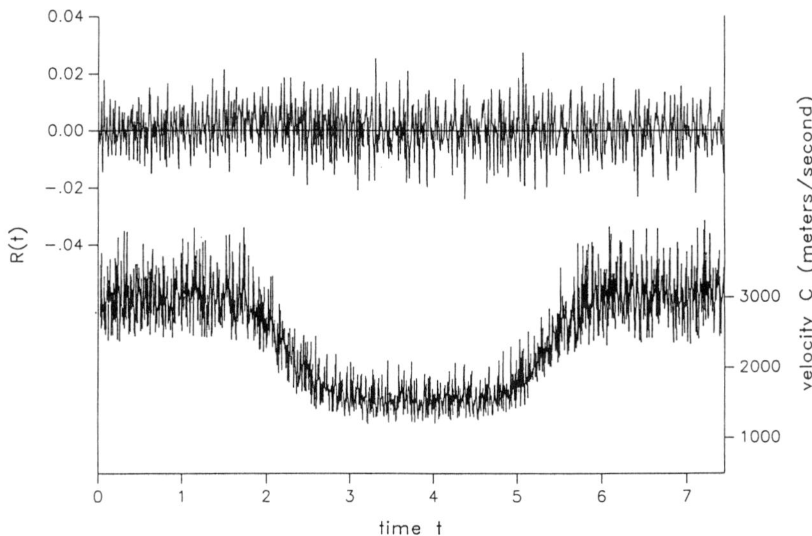

Figure 2. Backscatter and medium propagation speed ϵ = .02, maximum noise = 30%

The lower curve in Fig. 1 is the medium propagation speed as a function of depth. It clearly shows macroscopic variation with random fluctuations on the microscale. The upper curves are the acoustic velocity u(t,x) for different values of t. The pulse shape is clearly visible on the leading wavefront. However a long tail is left behind because of multiple scattering from the microstructure. The subsequent backscatter at x=0 is pictured as the upper curve in Fig. 2. For comparison the medium propagation speed is drawn as a function of travel time as the lower curve in Fig. 2.

In Fig. 3 we verify the Gaussian distribution of the backscatter at two times, t=.5 and t=.75. Since 2000 realizations were necessary to generate the pictured histograms, we raised the value of ϵ to ϵ=.05 for this run, to save computer time. We also used a constant mean speed profile, and a Gaussian pulse.

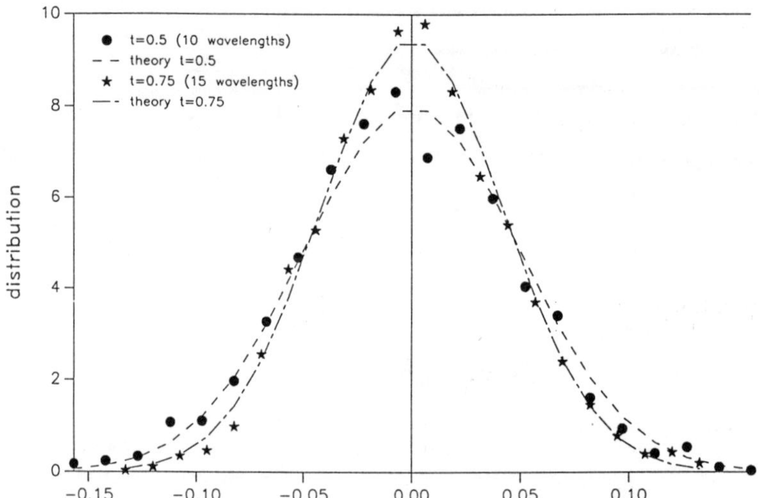

Figure 3. Probability distribution of the backscatter ϵ=.05, 2000 realizations

In Fig. 4 we verify the power spectrum prediction (18) for a window centered at t=.75, using a spectrum estimated from 1000 realizations of the random process. Pictured are the Gaussian pulse spectrum, f(ω), and a comparison of the predicted with the estimated spectra. The agreement is very good. The attempt to recover $\mu(t,\omega)$ by dividing the spectral estimate by $|f(\omega)|^2$ works reasonably well for frequencies within the support of f.

Finally, in Fig. 5 we compare the estimates of $\mu(t,\omega)$ with the theoretical curve for three different times. These may be plotted on the same curve by using the scaling property of μ, from (18)

$$\mu(t,\omega) = \frac{1}{t} \; \bar{\mu}(\omega \, t^{1/2}) \tag{19}$$

where

$$\bar{\mu}(\omega) = \frac{\alpha\omega^2}{(1+\alpha\omega^2)^2} \cdot \tag{20}$$

Overall the agreement is quite good, provided μ is estimated only at frequencies within the support of the pulse.

114

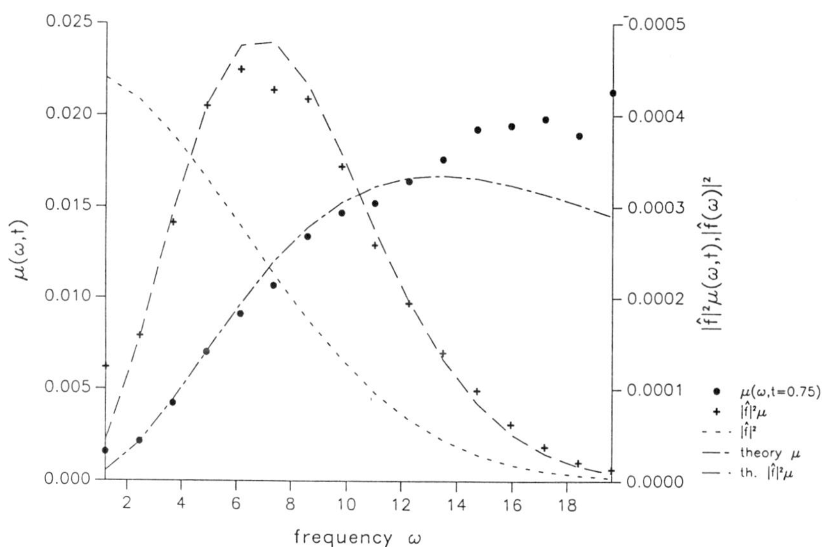

Figure 4. Local power spectral density $\epsilon=.05$, 1000 realizations

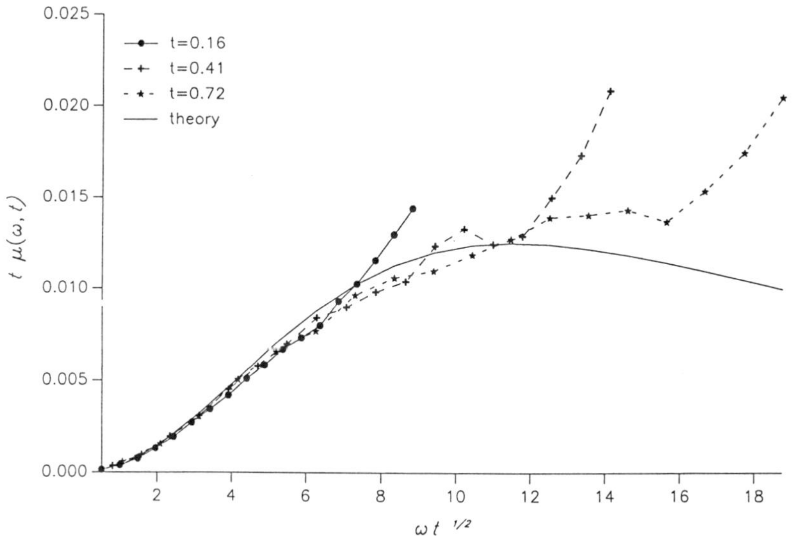

Figure 5. Scaled local power spectral density $\epsilon = .05$, 1000 realizations.

References

1. R. Burridge, G. Papanicolaou, P. Sheng and B. White, "Probing a random medium with a pulse", SIAM J. Appl. Math. 49, (1989) to appear. An abridged version appeared in "Non-Classical Continuum Mechanics", J. R. Knopps and A. A. Lacey eds, Cambridge Univ. Press (1987), pp 3-21.

2. R. Burridge, G. Papanicolaou, and B. White, "Statistics for pulse reflection from a randomly layered medium", SIAM J. Appl. Math. 47, (1987) pp.146-168.

3. P. Sheng, Z.-Q. Zhang, B. White, and G. Papanicolaou, "Multiple scattering noise in one dimension: universality through localization length scales", Phys. Rev. Letts. 57, (1986) pp. 1000-1003.

4. B. White, P. Sheng, Z.-Q. Zhang, and G. Papanicolaou, "Wave localization characteristics in the time domain", Phys. Rev. Letts. 59, (1987) pp. 1918-1921.

5. M. Asch, G. Papanicolaou, M. Postel, P. Sheng, and B. White, "Frequency content of randomly scattered signals I", Wave Motion, to be submitted.

6. G. Papanicolaou, M. Postel, P. Sheng, and B. White, "Frequency content of randomly scattered signals II: Inversion", Wave Motion, to be submitted.

7. A. Bayliss, K. Jordan, B. LeMesurier, and E. Turkel, "A fourth-order accurate finite-difference scheme for the computation of elastic waves", Bull. of the Seismol. Soc. of Am. 76, (1986) pp. 1115-1132.

Coherent Backscattering and Anderson Localization of Light

F.C. MacKintosh and Sajeev John

Physics Department, Princeton University, P.O. Box 708,
Princeton, NJ 08544, USA

1. Introduction

The propagation of classical waves such as light in strongly multiple scattering media has been
the subject of renewed interest. One motivation for this interest has been the suggestion that
Anderson localization might be observed in such systems [1,2]. Coherent backscattering of waves in
a random medium is the precursor to Anderson localization. The enhanced backward scattering in
highly disordered materials is responsible for the renormalization of the energy diffusion coefficient
as described by the scaling theory of localization [3]. In moderately disordered materials, in
which localization is not evident, coherent backscattering has nevertheless been observed. This
phenomenon of *weak* localization has been discussed in the context of electrons in metals by
KHMEL'NITSKII [4] and BERGMANN [5]. A number of experiments beginning in 1984 with
KUGA and ISHIMARU [6] have demonstrated the coherent backscattering of light [7-10]. The
coherent backscattering enhancement was discussed as early as 1969 for scattered radar waves [11].
GOLUBENTSEV [12] calculated the enhancement of the albedo for retroreflectance of scalar
waves from a random collection of point-like scattering centers. AKKERMANS, WOLF and
MAYNARD [13] obtained a calculated line shape of the backscattered intensity for scalar waves as
a function of scattering angle which agreed qualitatively with the experimental data. Although the
vector nature of the electromagnetic field does not alter the localization critical point, polarization
effects are apparent in angle resolved studies of the backscattered intensity [6-8]. STEPHEN and
CWILICH [14] have demonstrated theoretically that for linearly polarized light incident on a
scattering medium, the backscattering peak consists of a sharp narrow peak polarized parallel to
the incident light as well as a broader depolarized peak.

In this work we shall discuss the physical origins of coherent backscattering and the observed
polarization dependence, as well as the effects of finite geometry [14] and absorption in the medium.
We shall also describe in simple terms the results of the calculations in [15] of the backscattering
line shape in the presence of broken time-reversal and parity symmetries. Qualitatively each of
these effects leads to a suppression of the coherent intensity of the scattered light. In a dissipative
medium for example, the contribution of light scattered through long paths will be suppressed
due to the absorption of photons. In magneto-optically active materials, broken time-reversal
symmetry is manifest in unequal propagation speed for right and left circularly polarized light
traveling parallel to a strong magnetic field. Since the coherence of the long scattering paths
responsible for the sharp backscattering peak depends upon time-reversal symmetry, the effect
of a strong magnetic field is to round off this peak. This reduction of the coherent intensity is
apparent in the scattered light of the same linear or circular polarization as the incident light. The
latter case we refer to as the *helicity preserving* component of the backscattered light. A study
of the suppression of interfering diffusion paths as a function of external magnetic field would
be of particular interest near the mobility edge, where the absence of time-reversal symmetry
can lead to new critical exponents for the localization transition. In contrast, natural optical
activity has no effect on the helicity preserving component of the scattered light. This is because
right and left hand circularly polarized photons experience different dielectric constants in such
materials, independent of their direction of propagation. Time-reversal symmetry is maintained.

The breakdown of parity, however, leads to a decrease in the opposite helicity component of the backscattering peak, in which case the incident and backscattered light are related by a mirror symmetry.

Section 2 contains a discussion in simple physical terms of the results which are described in detail in [15]. The calculations are based on Green's functions for the linear wave equation with a spatially random dielectric function $\epsilon(\vec{r})$ [12,14], which is assumed to be uncorrelated on all finite length scales. This is the *white noise* approximation, valid for isotropic scatterers which are much smaller than a wavelength. Only the leading correction to the classical diffusion propagator has been considered. In perturbation theory, ordinary diffusion comes from the sum of *ladder* diagrams, whereas the leading correction corresponds to the sum of *maximally crossed* diagrams. This correction gives rise to a backscattering peak with angular width proportional to the small dimensionless parameter $\frac{\lambda}{2\pi l}$, where λ is the wavelength of light and l is the mean-free-path.

Qualitatively, we may say that localization occurs when a large fraction of the reflected waves are *coherently* backscattered. More precisely, the transition to localized states in dimensions $d > 2$ occurs when the ratio $\frac{\lambda}{2\pi l}$ becomes of order unity. This is the Ioffe-Regel condition [16], which is difficult to achieve in random dielectric materials. We shall discuss possible avenues for the observation of localized light, in which spatial correlations play a key role [17].

2. The Physical Picture of Coherent Backscattering

The classical multiple scattering treatment of wave propagation ignores phase correlation on length scales longer than the mean-free-path l. For elastic scattering in samples with dimensions much larger than l this treatment yields the diffusion equation with constant $D = \frac{lc}{3}$, where c is the wave velocity. For weak scattering ($l \gg \lambda$) this approximation is good except in the backward direction, where phase correlation of even long paths cannot be ignored. A variety of authors have discussed the significance of this interference [11,13]. Consider a half space of randomly placed scatterers and a typical path γ shown as the solid line in Fig. 1. Incident light with wave vector $\vec{k}_i = \vec{k}_0$ is scattered at the points $\vec{x}_1, \ldots, \vec{x}_N$ ($N > 1$) into intermediate states with wave vectors $\vec{k}_1, \ldots, \vec{k}_{N-1}$, and finally to the state $\vec{k}_N = \vec{k}_f$. For scalar waves undergoing an identical set of wave vector transfers, the scattering amplitudes at the points $\vec{x}_1, \ldots, \vec{x}_N$ are the same for the path γ and the time-reversed path $-\gamma$ (the dashed line). The interference between the waves scattered by these paths is determined solely by the difference in their optical path lengths. Thus the amplitudes for the two processes differ only by a phase $e^{i\vec{q}\cdot(\vec{x}_N - \vec{x}_1)}$, where $\vec{q} \equiv \vec{k}_i + \vec{k}_f$. For strict backward scattering ($\vec{q} = 0$) the amplitudes for the two paths are equal ($A_\gamma = A_{-\gamma}$) and hence the intensity will have a contribution $\sim |A_\gamma + A_{-\gamma}|^2 = 4|A_\gamma|^2$, which is twice the contribution $\sim |A_\gamma|^2 + |A_{-\gamma}|^2$ obtained when correlation is ignored. Thus a peak of height twice the classical intensity is obtained in the backward direction.

This peak has an angular width of $\Delta\theta \sim \frac{\lambda}{2\pi l}$ since, for angles θ larger than this, the argument of the phase $\vec{q} \cdot (\vec{x}_N - \vec{x}_1)$ is greater than unity. Consider the case of a wave incident normal to the interface and $\vec{x}_N - \vec{x}_1$ parallel to the interface. If the angle between $-\vec{k}_i$ and $\vec{k}_f$ is θ, then the coherence condition for small angles becomes $\vec{q} \cdot (\vec{x}_N - \vec{x}_1) = 2\pi\theta\frac{|\vec{x}_N - \vec{x}_1|}{\lambda} < 1$. In the diffusion approximation $|\vec{x}_N - \vec{x}_1|^2 \simeq D(t_N - t_1) \simeq \frac{lL}{3}$, where L is the total length of path γ. Thus typical paths of length L contribute only for angles less than $\theta_m = \frac{\lambda}{2\pi\sqrt{lL/3}}$. To facilitate a generalization to a propagating vector field it is useful to look at this in another way. Consider the path in k-space shown in Fig. 2, with wave vector transfers $\vec{g}_j = \vec{k}_j - \vec{k}_{j-1}$ ($j = 1, \ldots, N$). Equality of the individual scattering amplitudes at $\vec{x}_1, \ldots, \vec{x}_N$ requires that the reversed path be given by the transfers $\vec{g}_j$ in the reverse order: $\vec{g}_N, \ldots, \vec{g}_1$. The intermediate states no longer lie on the energy shell which is smeared by an amount $\hbar/\tau$, where $\tau = l/c$ is the mean lifetime of a plane wave in

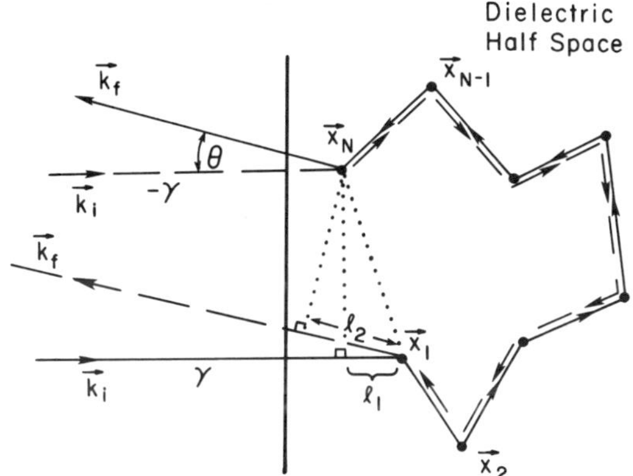

Fig. 1. A typical scattering path (γ) in real space and its time reversed path $(-\gamma)$ with which it coherently interferes for small angles θ. The relative phase between γ and $-\gamma$ is simply proportional to the optical path length difference $\ell_2 - \ell_1$, where $\ell_1 = (\vec{x}_1 - \vec{x}_N) \cdot \hat{k}_i$ and $\ell_2 = (\vec{x}_N - \vec{x}_1) \cdot \hat{k}_f$.

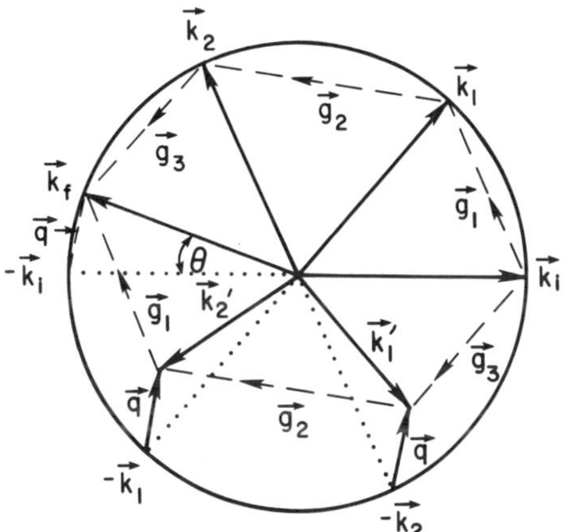

Fig. 2. A typical scattering path drawn in momentum space. Equality of the scattering amplitudes for the time reversed path requires that the wave vector transfers $\vec{g}_j$ be the same but in the reversed order. Corresponding plane wave states differ in their frequencies by an amount $\omega_{N-j} - \omega'_j \simeq c\vec{q} \cdot \hat{k}_j$, which leads to a phase difference $(\Delta\phi)_{\mathrm{rms}} \simeq \sqrt{N/3}\,q\ell$.

the disordered medium. For small $\vec{q}$, corresponding intermediate states differ in frequency by an amount $\omega_{N-j} - \omega'_j \simeq c\vec{q} \cdot \hat{k}_{N-j}$ $(j = 1,\ldots,N-1)$, where $\hat{k}_{N-j}$ are unit vectors in the direction of propagation of the intermediate plane wave states. The phase difference $\Delta\phi$ between paths is then

$$\Delta\phi = \sum_{j=1}^{N-1} c\tau_j \vec{q} \cdot \hat{k}_j \simeq q\ell \sum_{j=1}^{N-1} \cos\theta_j \,, \tag{1}$$

where τ_j is the lifetime of the plane wave state j. The values of $\cos\theta$ are random and hence the sum corresponds to a random walk. After averaging over all possible N step walks, the root-mean-square phase difference $(\Delta\phi)_{\mathrm{rms}}$ is approximately $\sqrt{\tfrac{1}{3}N}\,q\ell$, and the condition for coherence is the same as above (with $L \simeq N\ell$). In particular, the observed cusp near $\theta = 0$ depends on paths of arbitrary length. We thus expect that if coherence were limited to paths shorter than some maximum L_{m}, then the peak should be rounded off for angles less than $\dfrac{\lambda}{2\pi\sqrt{\ell L_{\mathrm{m}}/3}}$. Each of the symmetry breaking effects considered below corresponds qualitatively to such a path length cutoff.

One such example has been considered by STEPHEN and CWILICH [14] in which scattering is confined to a slab of finite thickness W. In this case paths of length greater than $L_{\mathrm{m}} = \frac{3W^2}{\ell}$ would have an appreciable probability of diffusing through the slab and being transmitted. Hence the portion of the reflected peak depending on paths longer than L_{m} would no longer be observed. The calculation described in [14] and [15] gives the line shapes shown in Fig. 3(a). These curves are consistent both with experiments [18] and with the qualitative arguments above. Small deviations from the calculated line shapes are expected for large angles and for very thin slabs, where only short random walks contribute. This can be understood as a breakdown of the diffusion approximation for short paths.

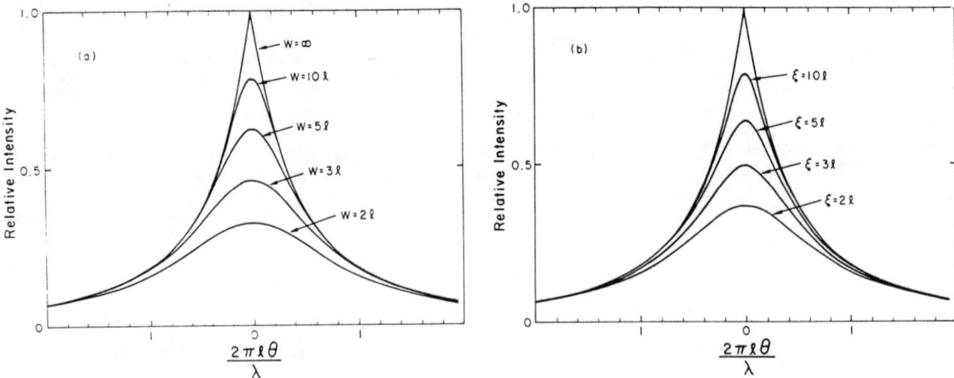

Fig. 3. The calculated scalar coherent backscattering line shapes. (a) As the width W of the scattering medium is reduced the central portion of the peak is suppressed by the termination of long scattering paths. (b) Similarly the peak is reduced as absorption is introduced. The mean penetration depth $\xi = \sqrt{\ell\ell_i/3}$ decreases with stronger absorption. Curves (a) and (b) are quite similar with the correspondence $W \leftrightarrow \xi$. These curves represent the coherent peak from which the isotropic background has been subtracted. The curves have been normalized by the isotropic background intensity in an infinite non-absorbing medium. Thus, the coherent peak of unit relative intensity represents an actual enhancement factor of 2 in the total reflected intensity.

Consider now what happens in the presence of absorption. One would expect that only paths of length less than the inelastic mean-free-path ℓ_i would contribute. This is similar to the finite geometry case above with the correspondence $W \leftrightarrow \xi \equiv \sqrt{\frac{\ell\ell_i}{3}}$ (see Fig. 3(b)). A subtle distinction between these two cases arises for large angles as can be seen in the calculated line shapes. Short paths will not reach the far side of the slab, but will still be somewhat attenuated by absorption. Thus in the first case, the curves for finite W approach very closely that of $W = \infty$ for sufficiently large angles—more so than the curves for finite ξ. The length ξ may be viewed as the mean depth into the medium which a photon diffuses before being absorbed. The value of this length may be found as follows. Diffusion in the presence of absorption is described by a momentum space propagator

$$\mathbf{D}(k,\omega) \equiv \frac{1}{D(k^2 + \frac{1}{\xi^2}) - i\omega} \, , \tag{2}$$

the steady state ($\omega = 0$) transform of which is

$$\mathbf{D}(r) = \frac{e^{-|r|/\xi}}{4\pi D r} \, . \tag{3}$$

Hence ξ has the interpretation of the *mean-penetration-depth* described above. On the other hand, the time dependent transform of (2) is

$$\mathbf{D}(r,t) = \frac{e^{-|r|^2/4Dt}}{(4\pi Dt)^{\frac{3}{2}}} e^{-Dt/\xi^2}. \tag{4}$$

Equation (4) describes the probability of finding a diffusing particle at a distance r from the point of origin after a time t. Thus a typical photon travels for a time $t_m \simeq \frac{\xi^2}{D} = \frac{3\xi^2}{\ell c}$ before being absorbed. It follows that $\xi = \sqrt{\frac{\ell\ell_i}{3}}$.

For a vector field propagating in a random medium, the condition that the optical path lengths be the same for reversed paths is necessary but no longer sufficient for constructive interference. For photons, scattering produces a sequence of rotations of the polarization vector, which simply ensure that the wave remains transverse. The properties of these rotations have been described by a number of authors [7,8,13]. We may describe these rotations of the polarization vectors $\hat{e}_j$ by matrices M_{j+1} which are real and symmetric. A sequence of scattering events rotates $\hat{e}_0$ by the product $M_N \cdots M_1$, and the reverse sequence by the product $M_1 \cdots M_N = (M_N \cdots M_1)^T$ for exact backward scattering. The matrices M_j do not commute for general $\vec{k}_j$ in three dimensions and only diagonal elements such as $\langle \hat{x}| M_N \cdots M_1 |\hat{x}\rangle = \langle \hat{x}| M_1 \cdots M_N |\hat{x}\rangle$ will remain coherent. Here, we have let $\vec{k}_i = k_0\hat{z}$ and $\vec{k}_f = -k_0\hat{z}$. Thus backscattered light polarized parallel to the incident light will retain the sharp peak associated with long paths, while the perpendicular component will be suppressed. This suppression is apparent in the central part of the peak, where long scattering sequences are important. The wings of the backscattering intensity, on the other hand, are dominated by shorter random walks where the non-commuting property of the rotation matrices is less serious. For example, backscattering from $\vec{k}_i$ to $-\vec{k}_i$ by an $N = 2$ step process can always be drawn in a single plane. Since rotations in this plane commute, coherence is maintained. Similarly, for incident circularly polarized light, coherence of long paths is retained only for reflected light of the same circular polarization. Consider right-hand circularly polarized incident light directed along the $\hat{z}$ axis and reflected light of the same polarization. The initial and final states may be described by: $|R_i\rangle \sim |x\rangle - i|y\rangle$ and $|R_f\rangle \sim |x\rangle + i|y\rangle$. Here too, the scattering amplitude is invariant under reversal: $\langle R_f| M_1 \ldots M_N |R_i\rangle = \langle R_f| M_N \ldots M_1 |R_i\rangle$. This helicity preserving channel corresponds to the experimental configuration used to remove single scattering events (which necessarily flip the helicity) from the observed intensity [18]. The incident light was circularly polarized and only scattered light of the same polarization was monitored. It should be noted that the backscattering peak in this channel closely resembles the calculated peak for scalar waves. (Compare Figs. 3(b) and 4(a) for example.) This similarity remains true in the presence of both confined geometry and absorption, for which we have also calculated the line shapes in the helicity preserving channel. It is shown in [15] that due to the elimination of single scattering contributions, the peak height in the helicity preserving channel is exactly twice the (classical) isotropic background intensity—as is the case for scalar waves. In contrast, these similarities to the scalar line shapes do not hold in the parallel linearly polarized channel [14]. The enhancement factor for the coherent peak in the parallel polarized channel is approximately 1.85. The vector corrections to the scalar line shape contribute unequally to the incoherent and coherent portions of the reflected light. In particular, single scattering events contribute to the incoherent intensity of parallel polarized light, but not to the coherent peak.

Associated with the polarization of the electromagnetic field are additional effects which break time-reversal symmetry in the absence of dissipation. One of these is the Faraday effect. The origin of this effect can be seen in a simple classical picture. When subject to a circularly polarized electromagnetic wave, the bound electrons execute circular orbits. A strong magnetic field ($\vec{B}$) along the direction of the wave will introduce a radial force on the electron. Depending on the sense of polarization of the wave, the orbit will either be increased or decreased—changing the dipole moment of the electron. Thus the effective dielectric constants for the two polarizations will differ. It is known that the refractive indices for the two polarizations are given by $n_{R/L} \simeq n_0 \mp \frac{\vec{g}\cdot\hat{k}}{2n_0}$, where $\vec{k}$ is the wave vector, $\vec{g} = f\vec{B}$ is the gyration vector and f is the Faraday constant [19]. Consider a path contributing to the helicity preserving channel with intermediate wave vectors $\vec{k}_j$ and helicities α_j (+1 for R and -1 for L). The helicities ($\alpha'_{N-j} = \alpha_j$) for the reversed path are chosen such that the amplitudes associated with each momentum transfer are the same. There is, however, an optical path length difference between the two paths with the introduction of a magnetic field. This phase difference is

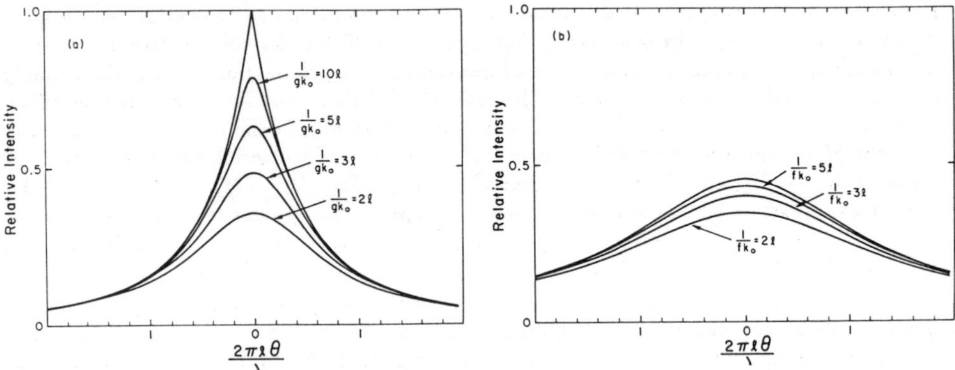

Fig. 4. (a) The effect of Faraday rotation on the peak line shape, where $\vec{g} = f\vec{B}$ is the gyration vector. The curves shown are for incident and reflected light of the same circular polarization. The opposite helicity channel is unaffected by a magnetic field. (b) The suppression of the intensity in the opposite helicity channel due to natural optical activity. The gyration vector $\vec{g} = f\hat{k}$ is proportional to the direction of propagation. The helicity preserving channel is unaffected since time-reversal symmetry is retained. In both cases the curves have been normalized by the background intensity in the helicity preserving channel.

$$\Delta\phi = ck_0 \sum \tau_j [\alpha_j \frac{\vec{g}\cdot\hat{k}_j}{2} - \alpha_j \frac{\vec{g}\cdot(-\hat{k}_j)}{2}] \simeq k_0 g\ell \sum \alpha_j \cos\theta_j \ , \tag{5}$$

where $k_0 = \omega/c$. For large N, α_j and $\cos\theta_j$ are uncorrelated. Hence the sum corresponds once again to a random walk, so that $(\Delta\phi)_{\mathrm{rms}} \simeq k_0 g\ell\sqrt{\frac{1}{3}N}$. Thus a magnetic field destroys coherence for such paths longer than $L_{\mathrm{m}} = \frac{1}{\ell(k_0 g)^2}$. This appears as a field dependent rounding off of the backscattering peak. On the other hand, for paths related by a mirror symmetry $\vec{k}'_{N-j} = -\vec{k}_j$ and $\alpha'_{N-j} = -\alpha_j$, the optical path lengths are the same. Faraday rotation does not affect the opposite helicity channel. As shown in [15], the effect of Faraday rotation on the helicity preserving portion of the backscattering peak is similar to that of absorption with the correspondence $\xi \leftrightarrow \frac{1}{gk_0}$. We also find that the reversed helicity channel is unaffected by Faraday rotation, as the above argument suggests. The results of this calculation are shown in Fig. 4(a) for the helicity preserving channel.

Natural optical activity provides a further example of a broken symmetry not previously considered in the context of localization: parity violation. Microscopically there is a difference in the dielectric constants for the two helicity states due to the electromagnetic response of helical molecules within the medium [20,21]. In contrast with Faraday rotation, the dielectric constants are independent of the direction of propagation. Some examples of optically active media are sugar, turpentine, selenium, tellurium, $AgGaS_2$, TeO_2 and quartz. The induced dipole moment $\vec{p} = \alpha\vec{E} - \beta\dot{\vec{H}}$ has the usual part proportional to the applied electric field $\vec{E}$ as well as a part opposite to the rate of change of the applied magnetic field $\vec{H}$. The latter contribution arises from electromotive forces in the helix given by Faraday's law of induction. The resulting constitutive relation for an optically active material yields an electric displacement vector $\vec{D} = \epsilon\vec{E} + i\vec{g}\times\vec{E}$ where $\vec{g} = f\hat{k}$ is the gyration vector parallel to the direction of photon propagation. For small f, the refractive indices for right and left hand polarizations are $n_{R/L} \simeq n_0 \mp \frac{f}{2n_0}$. For paths related by time-reversal in which $\alpha'_{N-j} = \alpha_j$, the optical path lengths for γ and $-\gamma$ are the same. Consequently, coherent backscattering into a helicity preserving state is unaffected by natural optical activity. For backscattering into a reversed helicity channel the incident and reflected states are related by a mirror symmetry. In this case, we arrive at a phase difference $(\Delta\phi)_{\rm rms} \simeq k_0 f \ell \sqrt{\frac{1}{3}N}$, which diminishes the backscattering intensity as the rotatory power f increases. This is illustrated in Fig. 4(b).

3. Anderson Localization of Light

As we have said, for sufficiently strong disorder the coherent backscattering of waves leads ultimately to the absence of diffusion, or the *strongly* localized regime. We may see this from a simple physical picture [4]. The diffusion propagator in (4) tells us that for all times t there is a finite probability (proportional to $t^{-d/2}$ in d dimensions) for a diffusing particle to return to its point of origin. A scalar wave following such a path, which begins and ends at the same point in space, will interfere constructively with the reversed path. This wave interference yields a probability of return to the point of origin which is twice as high as for a classical particle. This is the same interference responsible for coherent backscattering, and it results in a reduction in the classical transport of wave energy between widely separated points.

The experimental observation of coherent backscattering of electromagnetic waves suggests the possibility of observing *strong* localization of light. This would be of great interest for several reasons. First, photons would provide a non-interacting system in which to test the scaling theory of localization [3]. The situation for electrons is greatly complicated by the presence of strong interactions between the electrons [22]. Also the availability of high resolution probes such as tunable lasers and sensitive detectors should allow very precise studies of the localization critical region.

Implicit in the discussion above is the analogy between the wave equation satisfied by light and the Shrödinger equation

$$-\frac{\hbar^2}{2m}\nabla^2\psi(\vec{x}) + V'(\vec{x})\psi(\vec{x}) = \mathcal{E}\psi(\vec{x}) \tag{6}$$

satisfied by the wave function ψ of an individual electron moving in a random potential V'. The wave equation for the electric field $E(\vec{x})$ may be written as

$$-\nabla^2 E - k_0^2\epsilon'(x)E = k_0^2\epsilon_0 E \; , \tag{7}$$

where $k_0 \equiv \omega/c$ and ω is the frequency of the light. Here the dielectric constant of the medium is written as a constant part ϵ_0 plus a spatially fluctuating part $\epsilon'(\vec{x})$ [17]: $\epsilon(\vec{x}) = \epsilon_0 + \epsilon'(\vec{x})$. We have neglected a term $\nabla(\nabla \cdot E)$ in (7) and have treated the electric field as a scalar—in much the same way that we have implicitly neglected the *spin* of the electron in (6). In (7), $-k_0^2\epsilon'(\vec{x})$ plays the role of the random potential, while $k_0^2\epsilon_0$ appears as the "energy" of the particle. There are, however, important distinctions between the propagation of electromagnetic waves and electrons—even within the scalar approximation above. In dissipationless dielectric media, the dielectric constant $\epsilon(\vec{x})$ is everywhere real and positive—*i.e.* $\epsilon_0 > -\epsilon'$. This corresponds to the restriction $\mathcal{E} > V'$ in (6). This is the condition for a particle whose *classical* motion is unrestricted. This is sketched in Fig. 5. Thus in the geometric optics limit of wavelengths λ much smaller than the length of correlations within the medium, we expect that light will not be localized. Furthermore, for sufficiently strong fluctuations V', (6) leads (in dimensions $d > 2$) to a transition to localized states ψ for energies $\mathcal{E}$ below some critical value $\mathcal{E}_c$. In contrast, both the energy $k_0^2\epsilon_0$ as well as the strength of fluctuations $k_0^2\epsilon'$ in (7) decrease with the frequency of the light. For low frequencies (or wavelengths larger than the size, a, of scattering particles), we arrive at Rayleigh scattering, for which the cross section decreases with the fourth power of the frequency in three dimensions. In other words, the scattering mean-free-path grows as $\ell \sim \lambda^4$. In neither of the extremes of

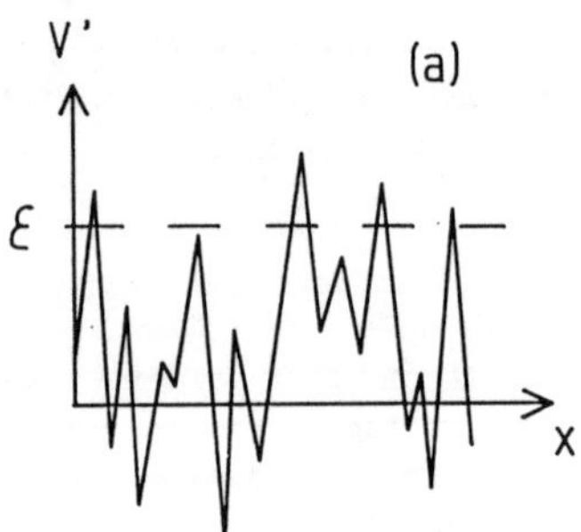
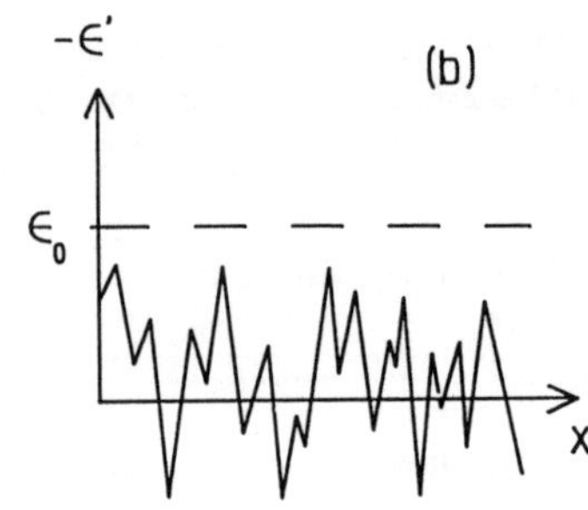

Fig. 5. (a) The random potential V' in which the electrons move. If the fluctuations are strong enough, the electrons may become trapped *classically* in potential wells. In one and two dimensions, however, no extended *quantum* states are possible even for arbitrarily weak fluctuations. (b) In dissipationless dielectric media, the strict positivity of the dielectric constant requires that $\epsilon_0 > -\epsilon'$. Nevertheless, strong fluctuations ϵ' may lead to renormalized diffusion and a localization transition in dimensions greater than two.

high or low frequencies can we expect that light will be localized, for which it is required that $\ell \lesssim \lambda$ [1,16]. Thus if localization is to be observed, it will be seen within some (perhaps narrow) window of frequencies $\omega_1 < \omega < \omega_2$ [1]. This is sketched in Fig. 6, where the expected behavior of ℓ as a function of λ is shown. For large wavelengths ℓ grows rapidly with λ due to Rayleigh scattering, while for small wavelengths the mean-free-path is limited by the length scale a of correlations within the medium. This points out the importance of going beyond the *white noise* approximation, since we may expect localized states only when all three lengths are nearly the same: $\lambda \sim \ell \sim a$ [23].

It has also been suggested recently that large scale correlations might play an important role in the localization of light [17]. Again we return to the analogy with electronic systems. It is well known that electronic states near a band gap may become localized in the presence of only moderate disorder. This is due to the highly suppressed density of states near such a gap. For electron wave functions to become extended, they must overlap sufficiently with nearby states of nearly the same energy [24]. In a perfect crystal, a band gap due to Bragg reflections corresponds to a range of electron energies for which free propagation is forbidden. With the introduction of weak disorder, a band gap (or pseudo gap in amorphous materials) tends to fill in with localized states [25]. The Ioffe-Regel criterion [16], which has also been derived for photons [1], assumes a free particle density of states and thus applies only to the center of a band. Near a band edge, the group velocity of a wave packet approaches zero. The dimension λ' of the envelope of the packet

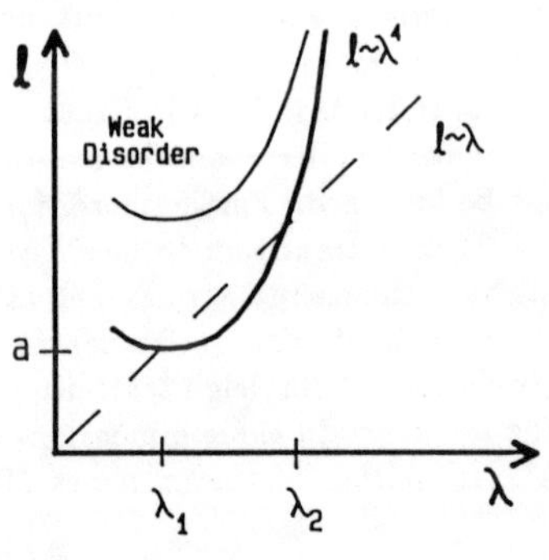

Fig. 6 The mean-free-path ℓ grows as λ^4 for long wavelengths due to Rayleigh scattering. In the geometric optics limit of short wavelengths, ℓ is at least as large as the characteristic size a of correlations within the medium. Between these two regimes, localization for strong scattering is characterized by two mobility edges at λ_1 and λ_2.

diverges. It is when this length λ' becomes large compared with ℓ that determines the mobility edge separating extended from localized states.

The existence of photonic band gaps has been reported by YABLONOVICH [26] for microwaves. Theoretical work [17,27] has shown that a photonic band gap can be created in an fcc lattice of dielectric spheres with volume filling fraction of approximately 11% and with a refractive index ratio of greater than 2.8 relative to the background. The expected gap in the density of states $\rho(\omega)$ is shown in Fig. 7(a). With a small degree of disorder, it may be possible to create localized photonic states near the band edges, as indicated by Fig. 7(b).

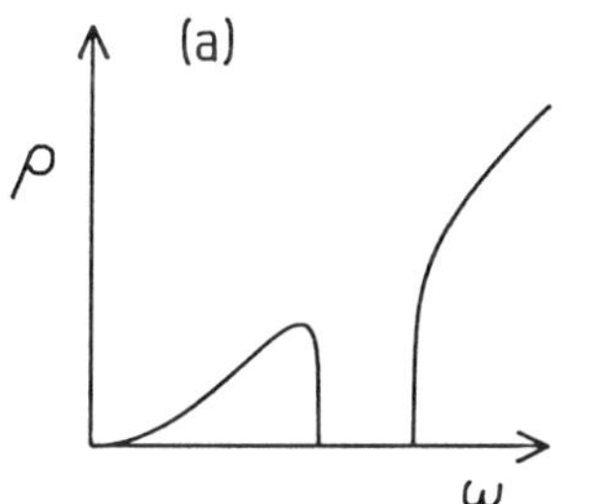
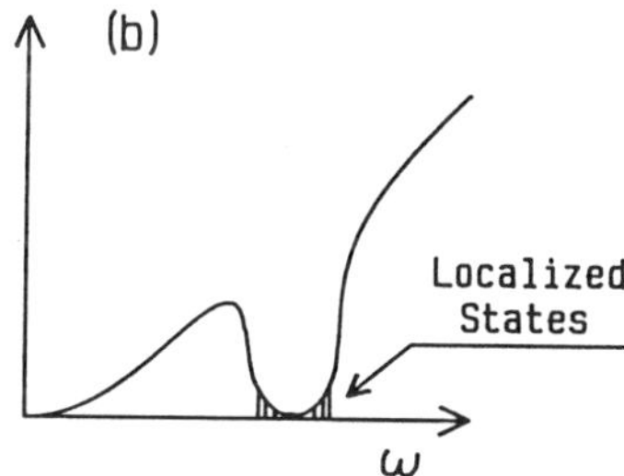

Fig. 7 (a) A photonic band gap in the absence of disorder is characterized by a density of states which falls rapidly to zero near the range of forbidden frequencies. (b) With disorder, localized states may develop near the band edges, where the density of states is small.

4. Discussion

The observation of coherent backscattering raises anew the prospect of observing Anderson localization of light. This would provide a non-interacting system in which to study the scaling theory of localization of classical waves. Faraday rotation and natural optical activity, in which are broken time-reversal and parity symmetries, have observable consequences for the coherent backscattering of light. The breakdown of these symmetries may lead to new critical exponents for the localization transition in dimensions $d > 2$.

It is expected that localization can occur only when the length scale of correlations within the scattering medium becomes comparable with the mean-free-path ℓ and the wavelength λ. Mie resonances for scattering particles of size $a \gtrsim \lambda$ are known to provide enhanced scattering, and have been suggested as a possible mechanism for the localization of light [28]. Furthermore, we expect that large scale correlations between scattering particles can lead to a gap in the photon density of states, and to localized states near the gap. This should be possible with a smaller degree of disorder than required by the Ioffe-Regel criterion, as derived in the *white noise* model [1].

References

1. S. John: Phys. Rev. Lett. **53**, 2169 (1984)
2. P. W. Anderson: Phil. Mag. B **52**, 505 (1985)
3. E. Abrahams, P. W. Anderson, D. C. Liccardello, T. V. Ramakrishnan: Phys. Rev. Lett. **42**, 673 (1979)
4. D. E. Khmel'nitskii: Physica **126B**, 235 (1984); B. L. Al'tshuler, A. G. Aronov, D. E. Khmel'nitskii, A. I. Larkin: In *Quantum Theory of Solids*, ed. by I. M. Lifshits (Mir, Moscow, 1982), pp. 130-237

5. G. Bergmann: Phys. Rep. **107**, 1 (1984)
6. Y. Kuga, A. Ishimaru: J. Opt. Soc. Am. A **1**, 831 (1984)
7. M. van Albada, A. Lagendijk: Phys. Rev. Lett. **55**, 2692 (1985)
8. P. E. Wolf, G. Maret: Phys. Rev. Lett. **55**, 2696 (1985)
9. S. Etemad, R. Thompson, M. J. Andrejco: Phys. Rev. Lett. **57**, 575 (1986)
10. M. Kaveh, M. Rosenbluh, I. Edrei, I. Freund: Phys. Rev. Lett. **57**, 2049 (1986)
11. K. M. Watson: J. Math. Phys. **10**, 688 (1969); D. A. de Wolf: IEEE Trans. Antennas Propag. **19**, 254 (1971)
12. A. A. Golubentsev: Zh. Exsp. Teor. Fiz. **86**, 47 (1984) (Soviet Phys. JETP **59**, 26 (1984))
13. E. Akkermans, P. E. Wolf, R. Maynard: Phys. Rev. Lett. **56**, 1471 (1986)
14. M. J. Stephen, G. Cwilich: Phys. Rev. B **34**, 7564 (1986)
15. F. C. MacKintosh, S. John: Phys. Rev. B **37**, 1884 (1988)
16. A. F. Ioffe, A. R. Regel: Prog. Semicond. **4**, 237 (1960)
17. S. John: Phys. Rev. Lett. **58**, 2486 (1987)
18. S. Etemad, R. Thompson, M. J. Andrejco, S. John, F. MacKintosh: Phys. Rev. Lett. **59** 1420 (1987)
19. L. D. Landau, E. M. Lifshitz: In *Electrodynamics of Continuous Media*, (Pergamon, London, 1960)
20. A. Yariv, P. Yeh: In *Optical Waves in Crystals*, (Wiley, New York, 1984)
21. E. U. Condon: Rev. Mod. Phys. **9**, 432 (1937)
22. P. A. Lee, T. V. Ramakrishnan: Rev. Mod. Phys. **57**, 287 (1985)
23. S. John, M. J. Stephen: Phys. Rev. B **28**, 6358 (1983)
24. P. W. Anderson: Phys. Rev. **109**, 1492 (1958)
25. J. M. Ziman: In *Models of Disorder*, (Cambridge Univ. Press, Cambridge, 1979)
26. E. Yablonovich: Bull. Am. Phys. Soc. **33**, 421 (1988)
27. S. John, R. Rangarajan: to be published
28. P. Sheng, Z. Zhang: Phys. Rev. Lett. **57**, 1879 (1986)

Weak Localization, Correlations and Fluctuations in Light Scattering from a Disordered Medium

M.J. Stephen

Physics Department, Rutgers University, Piscataway, NJ 08855, USA

1. INTRODUCTION

There has been a lot of work recently on the transport, localization and fluctuations of electrons in disordered metals.[1] Localization of electrons in a disordered metal is a one body problem and the statistics of the electrons is not important. What is important is that the electron behaves as a wave and localization results from the interference of the various multiply scattered waves. Thus the same phenomena which occur and methods which apply for electrons in disordered metals should be useful for other kinds of waves in disordered media e.g. phonons, spin waves, electromagnetic waves etc. The case of light scattered from a disordered medium is particularly interesting because experiments can easily be carried out and it is possible to observe the interference effects which give rise to localization. In this lecture we will discuss the theory briefly in Section 2, coherent back scattering in 3, intensity correlations and fluctuations in 4 and time dependent effects in 5.

2. THEORY

The basic equation that we begin with is the wave equation

$$[\nabla^2 + k^2(1+\epsilon(r))] \, E(r) = 0 \tag{2.1}$$

where $k = \frac{2\pi}{\lambda}$ is the wave vector of the wave, E is the wave field (either scalar or vector) and $\epsilon(r)$ is the random part of the dielectric constant. We have assumed it to be time independent so that the scattering is purely elastic (in Sect. 5 we discuss inelastic scattering) and we take ϵ to have zero mean and a correlation function of the spatial white noise form

$$\langle \epsilon(\vec{r}) \, \epsilon(\vec{r}') \rangle = \Delta \, \delta(\vec{r}-\vec{r}') \tag{2.2}$$

The important assumptions that we will make are (a) the scattering from a single impurity is of the Raleigh form i.e. the scatterers are much smaller than the wavelength of the light. This is consistent with the white noise assumption (2.2). (b) We are in the region of weak localization i.e. the mean free path of the wave is much larger than the wavelength. We can then use a form of perturbation theory to calculate the field correlation functions starting from (2.1). This perturbation theory is basically an expansion in ladder graphs, the small expansion parameter being $(k\ell)^{-1}$. The ladder graphs give rise to diffusive transport of the intensity in the medium. The mean free path is determined by the density of scatterers and in the weak scattering limit in $\ell = \frac{4\pi}{\Delta k^4}$. This is of the Rayleigh form and varies with the fourth power of the wavelength. The diffusion constant is $D = \frac{c\ell}{3}$ when c is the light velocity.

Springer Proceedings in Physics, Vol. 39   **Disorder and Nonlinearity**   127
Editor: A.R. Bishop   © Springer-Verlag Berlin, Heidelberg 1989

The coherent part of the light beam $\langle E(r) \rangle$ (angular brackets indicate an average over imparity configurations) will only penetrate a short distance of order ℓ into the medium but provides the source for the diffusely scattered light. Properties of the diffusely scattered light are contained in the correlation function (or coherence function)

$$\Gamma(\vec{R},\vec{r}) = \langle E^*(\vec{R}+\tfrac{\vec{r}}{2})\, E(\vec{R}-\tfrac{\vec{r}}{2}) \rangle_c \qquad (2.3)$$

where the subscript c indicates a cumulant. When we take the Fourier transform function on the variable $\vec{r}$ to form $\Gamma(\vec{R},\vec{q})$ this function is proportional to the average intensity of radiation at point $\vec{R}$ with wave vector $\vec{q}$ i.e. travelling in the direction $\vec{q}$. As the scattering is elastic $q{\sim}k$ and it is convenient to integrate over q giving the total intensity of radiation at $\vec{R}$ travelling in the direction of the unit vector $\vec{s}$ as

$$J(R,\vec{s}) = \int_0^\infty q^2 dq\, \Gamma(R,q\vec{s}) \qquad (2.4)$$

The correlation function (2.3) is one of the quantities of interest.

3. COHERENT BACK SCATTERING

The diffusely scattered light from the disordered medium is the result of many multiply scattered waves. The total wave amplitude of the scattered light, E_s, is the sum of the amplitudes a_α of the different scattered waves which we label by α i.e. $E_s = \sum_\alpha a_\alpha$. The average intensity of the scattered light is then

$$\langle |E_s| \rangle = \sum_\alpha \langle |a_\alpha|^2 \rangle + \sum_{\alpha \neq \beta} \langle a_\alpha^* a_\beta \rangle \qquad (3.1)$$

As the medium is disordered it is generally assumed that the different amplitudes a_α have phases which are completely uncorrelated in which case the second term of (3.1) averages to zero. This is what is done in the usual treatment of Rayleigh scattering from a disordered medium. Interference usually results from some symmetry and in the case where the scattering is elastic i.e. the impurities are stationary, the only global symmetry that the disordered medium possesses is that of time reversal. It is now well known that two multiply scattered waves following time reversed paths interfere constructively in the backward direction. This is true for all multiple scattering paths of two or more scatterings. This simple argument shows that the intensity reflected in the backward direction is increased by a factor 2 over the background i.e. the terms $\langle a_\alpha^* a_\beta \rangle = \langle a_\alpha^* a_\alpha \rangle$ if β is the time reverse of path α (for backscattering). The earliest reference to this effect of which I am aware is in a footnote to a paper by Watson[2] where it is attributed to Ruffine. It was also discussed in 1966 by Langer and Neal[3] for electrons. It has now been observed experimentally by a number of groups[4-8] and called coherent backscattering (cbs).

The width of the backscattered peak can be determined by a simple argument. Consider a diffusive multiple scattering path of linear dimension L. If the scattered wave and the time reversed scattered wave are emitted at a small angle θ to the backward direction they have a path difference $L \sin \theta$ and interfere constructively provided $L \sin \theta < \lambda$. The shortest diffusion path has $L{\sim}\ell$ and thus constructive interference only occurs if $\theta < \lambda/\ell$. The angular line shape has been given by a number of authors for scalar waves[9-11] and is

$$J(\theta) \sim 1 + \frac{1}{(1+k\ell\sin\theta)^2}$$

where the first term represents the Rayleigh scattering and the second term the cbs. The line shape is triangular for θ small.

The above argument indicates that cbs at small angles comes from long diffusion paths L. Thus any perturbation which cuts off the long diffusion paths will give rise to a rounding of the triangular peak e.g. finite size effects and absorption[10,11]. Similarly any effects which break the time reversal symmetry of the scattering medium will also lead to a rounding and reduction in the cbs peak e.g. diffusion or motion of the scatterers[9], applied magnetic fields (Faraday effect) and optical activity.[12]

Up to now we have only considered scalar waves and it is found that the effects of polarization are interesting.[11] It is not difficult to show[12] that only the polarization preserving scattering shows true cbs. After a single scattering of light with polarization $\vec{P}$ from $\vec{k}$ to $\vec{k}'$ the new polarization vector is $\vec{P}' = \vec{k}' \times (\vec{k}' \times \vec{P})$. This is conveniently written in matrix form

$$P' = (M(k')) P$$

when M is a symmetric matrix. The polarization of a wave multiply scattered n times, P_n^+ and that of the time reversed wave P_n^- are given by

$$P_n^+ = AP, \quad P_n^- = \tilde{A} P$$

when $A = M(k_n) \ldots M(k_2) M(k_1)$ and $\tilde{A}$ is the transpose of A. If the initial polarization $P = \binom{0}{1}$ then

$$P_n^+ = \begin{pmatrix} A_{13} \\ A_{23} \\ A_{33} \end{pmatrix} \quad P_n^- = \begin{pmatrix} A_{31} \\ A_{32} \\ A_{33} \end{pmatrix}$$

so that if the polarization is preserved the two waves are in phase while because $A_{13} = A_{31}$, $A_{23} = A_{32}$ the two waves with rotated polarization in general do not interfere constructively. A more detailed theory[11] shows that for polarized waves the cbs peak in the polarizing preserving channel is reduced in magnitude so that we no longer get exactly twice the intensity in exactly the backward direction that was found for scalar waves. In addition several broader (in angle) peaks of approximate Lorentzian shape occur in the back scattering direction both for the polarizing preserving channel and for the channel corresponding to the rotated polarization. These results are in good accord with the experiments.

4. INTENSITY FLUCTUATIONS AND CORRELATIONS

The speckle patterns which are observed when coherent light is scattered from a disordered surface or medium result from the large intensity fluctuations of the scattered light.[14] The gross features of a speckle pattern can be understood by again regarding the total wave amplitude of the scattered light, E_s, as the sum of many independent amplitudes a_α as in Sect. 3. Neglecting any correlations between these amplitudes the average intensity is $\langle I \rangle = \sum_\alpha \langle |a_\alpha|^2 \rangle$ and the intensity fluctuations

$$\langle I^2 \rangle = \sum_{\alpha\beta\gamma\delta} \langle a_\alpha^* a_\beta a_\gamma^* a_\delta \rangle = 2\langle I \rangle^2 \tag{4.1}$$

when the factor 2 comes from the two possible ways of identifying the subscripts i.e. $\alpha=\beta$, $\gamma=\delta$ or $\alpha=\delta$, $\beta=\gamma$ in (4.1). The fluctuations in the intensity are thus as large as the average intensity. This argument can be extended to show that the distribution of the intensity $P(I) = \frac{1}{\langle I \rangle} e^{-I/\langle I \rangle}$, the Raleigh or exponential distribution. There do not appear to be any interesting corrections to this law (at least in the weak scattering limit) and it is well obeyed experimentally.

We now turn to the intensity correlation function $C(R_{12}) = \langle I(R_1)I(R_2) \rangle_c$ defined as a cumulant. The simple theory of this correlation function neglecting correlations between different multiply scattered waves was worked out by Shapiro[15] who showed that

$$C_o(R_{12}) = \langle I(R_1) \rangle \langle I(R_2) \rangle \left(\frac{\sin kR_{12}}{kR_{12}}\right)^2 e^{-R_{12}/\ell} \tag{4.2}$$

when ℓ is the mean free path and the subscript zero means the uncorrelated approximation to C. This reduces to (4.1) (without the 2 because C is a cumulant) when $R_1=R_2$ and shows that the correlations decrease exponentially in a mean free path. In addition the speckle pattern has structure on the scale of the wavelength.

However, we expect that (4.2) is not correct when $R_{12} \gg \ell$ because the transport of the light is diffusive in nature and diffusion is a long range phenomenon. In addition the diffusing intensity has an underlying wave field so that interference can occur. These interference effects were shown by Stephen and Cwilich[16] to lead to long range correlations of the intensity decreasing like R_{12}^{-3} for $R_{12} \gg \ell$. Their result for the correlation function to order $(k\ell)^{-2}$ in perturbation theory is

$$C_r(R_{12}) = C_o(R_{12}) + \frac{1}{k^2\ell^2} \left(\frac{\ell}{R_{12}}\right)^3$$

$$C_t(R_{12}) = C_o(R_{12}) + \frac{1}{k^2\ell^2} \left(\frac{\ell}{L}\right)^3 F(R_{12}/L) \tag{4.3}$$

when the subscript r and t indicate the correlation for the light diffusely reflected and transmitted from a slab of thickness L respectively. The function $F(R/L) = L/R-1...$ for $R<L$ and $F(R/L) = e^{-R/L}$ for $R>L$. The second terms in (4.3) are only valid if $R_{12} \gg \ell$. Thus the intensity correlation function drops off exponentially for $R_{12} \sim \ell$ but has a power law tail for $R_{12} \gg \ell$. The size of this tail is reduced by the factor $(k\ell)^{-2}$. The first and second term in (4.3) join up reasonably smoothly for $R_{12} \sim \ell$. These long range correlations have not yet been observed presumably because they are quite small.

Other quantities of interest are the reflection and transmission fluctuations. For unit incident intensity on a slab of cross sectional area A and thickness L we define the reflection and transmission coefficients r and t by

$$r = \frac{1}{A} \int d^2R \, I_r(R)$$

$$t = \frac{1}{A} \int d^2R \, I_t(R) \tag{4.4}$$

where the integrals extend over the reflecting surface and transmitting surfaces respectively and I_r and I_t are the diffusely reflected and transmitted intensities on these surfaces respectively. The average reflection coefficient $\langle r \rangle \sim 1$ for a thick slab and $\langle t \rangle \sim \ell/L$ the dependence on $1/L$ resulting because the transport is diffusive. The fluctuations in r and t can be calculated from the correlation functions (4.3) and give

$$\langle r^2 \rangle_c \sim \frac{1}{k^2\ell^2} \left(\frac{\ell^2}{A} \right)$$

$$\langle t^2 \rangle_c \sim \frac{1}{k^2\ell^2} \left(\frac{\ell^2}{A} \right) \left[\left(\frac{\ell}{L} \right)^2 + \left(\frac{\ell}{L} \right) \right] \tag{4.5}$$

In the case of reflection both terms in (4.3) give contributions to $\langle r^2 \rangle_c$ of the same order or magnitude and have been combined in (4.5). For transmission fluctuations the long range correlations give rise to the second term in (4.5) which is larger than the first term by a factor $L/\ell \gg 1$. Thus an alternative experimental approach is to measure the transmission fluctuations which are larger (by L/ℓ) than predicted by simple theory.

Analogous problems to those considered here arise in the conductance and its fluctuations of electrons in disordered metals and there has been considerable experimental and theoretical interest[17,18] recently in this area. The relation of transmission fluctuations to conductance fluctuations has been discussed by Feng, Kane, Lee and Stone.[19]

5. TEMPORAL FLUCTUATIONS

In this section we discuss how the dynamical properties of the scattering medium determine the spectral properties of the scattered intensity. Often it is assumed that the scattering is weak and can be treated in Born approximation. In this case the scattering is directly related to the dynamical structure factor of the medium. Here we are interested in the case where the wave is multiply scattered in the medium which will occur if the mean free path is much less than its dimensions. Owing to the multiple scattering, the scattered intensity will not depend in an important way on the scattering angle but its spectral properties will be determined by the dynamics of the medium. The total scattered intensity is obtained by summing the contribution of all possible multiple scattering paths. The relaxation time depends on the number of scattering events i.e. on the length of the multiple scattering path. The total intensity thus contains a broad range of relaxation times.

In this case the dielectric constant in (2.1) varies slowly with the time because of the motion of the scatterers and we take its correlation function to be

$$\langle \epsilon(r,t) \rangle \, \epsilon(r^1 t^1) \rangle = C(r-r^1, \, t-t^1) \tag{5.1}$$

It is useful to consider some simple examples:
(a) The scatterers diffuse with a diffusion constant D_i. The Fourier transform of C is

$$C(q,t) = \Delta e^{-q^2 D_i |t|} \tag{5.2}$$

(b) The scatterers have a mass m and a Maxwell-Boltzmann velocity distribution

$$C(q,t) = \Delta\, e^{-q^2 t^2 / 2m\beta} \tag{5.3}$$

where $\beta = 1/kT$.

The condition that the variation in ϵ be slow means that we only consider times $t < \tau_\lambda$ where τ_λ is the time for a scatterer to move a wavelength. In the above two examples $\tau_\lambda = (4D_i k^2)^{-1}$ and $\tau_\lambda = (m\beta/2k^2)^{1/2}$. The frequency change on scattering is small and we can consider the propagation of an almost monochromatic wave with frequency close to $\omega = ck$. The mean free path of the wave ℓ will not depend on the motion of the scatterers and is determined by the density of scatterers, $\ell = 4\pi/\Delta k^4$.

As a simple example we consider a monochromatic point source of light at the origin in an infinite medium and examine the spectral properties of the scattered intensity at point R. In particular we consider the correlation function

$$\Gamma(R,t) = \langle E(R,t)\, E^*(R,o)\rangle_c \tag{5.4}$$

Consider a scattering path of length L which goes from the origin to R. At each scattering event we get a Doppler shift $\Delta\omega_i$ and the total Doppler shift is $\Omega_L = \sum_i \Delta\omega_i$ when i runs from 0 to L/ℓ. The contribution to (5.4) from this path is proportional to

$$\langle e^{i\Omega_L t}\rangle = (\langle e^{i\Delta\omega t}\rangle)^{L/\ell} = (f(t))^{L/\ell}$$

where f is the angular average of the correlation function (5.1)

$$f(t) = \frac{1}{4\pi\Delta} \int d\vec{s}^{\,1}\, C(k(\vec{s}-\vec{s}^{\,1}),t) \tag{5.5}$$

where $\vec{s}$ and $\vec{s}^{\,1}$ are unit vectors. In the two examples (5.2) and (5.3)

$$f^{(a)}(t) = \frac{\tau_\lambda}{|t|}(1-e^{-|t|/\tau_\lambda})$$

$$f^{(b)}(t) = \frac{\tau_\lambda^2}{t^2}(1-e^{-t^2/\tau_\lambda^2}) \tag{5.6}$$

We now assume that the light diffuses in the medium and then the probability of reaching R after L steps is proportional to $(\ell/L)^{3/2} e^{-3R^2/4\ell L}$. We now average (5.5) with respect to this distribution

$$\Gamma(t) \sim \int_0^\infty dL (\frac{\ell}{L})^{3/2}\, e^{-3k^2/4\ell L}\, (f(t))^{L/\ell} \tag{5.7}$$

132

The integrand has a maximum at $L^2 = 3R^2/4 \log \frac{1}{f}$ which gives

$$\Gamma(t) \sim e^{-\frac{\sqrt{3}R}{\ell}(\log \frac{1}{f})^{1/2}} \tag{5.8}$$

In the two cases (5.7) we thus find

$$\Gamma_a(t) \sim e^{-\frac{R}{\ell}\frac{3t}{(2\tau_\lambda)}^{1/2}}$$

$$\Gamma_b(t) \sim e^{-(\frac{3}{2})^{1/2}\frac{Rt}{\ell\tau_\lambda}} \tag{5.9}$$

The important features of these results are that we get a broad distribution of relaxation times which depends on the path length R of the scattered wave. For the case of diffusing scatterers the time correlation function decays as a stretched exponential and for a free gas it decays exponentially. They are very different from the case of single scattering, (5.2) and (5.3). In the multiple scattering case it is still possible to obtain information about the dynamics of the medium τ_λ. Results for different geometries have been given by Stephen[20].

A number of experimental studies of the time and frequency dependence of the field correlation function in the multiple scattering case have been carried out[21-23] generally in reasonable agreement with these results.

REFERENCES

1. For a recent survey see "Localization, Interaction and Transport Phenomena in Impure Metals", G. Bergmann, Y. Bruynseradi and B. Kramer, Editors (Springer-Verlag, New York 1985).
2. K. M. Watson, J. Math. Phys. 10, 688 (1969).
3. J. S. Langer and T. Neal, Phys. Rev. Lett. 16, 984 (1966).
4. M. P. van Albada and A. Lagendijk, Phys. Rev. Lett. 55, 2692 (1985).
5. P. E. Wolf and G. Maret, Phys. Rev. Lett. 55, 2696 (1985).
6. S. Etemad, R. Thompson and H. J. Andrejco, Phys. Rev. Lett. 57, 575 (1986).
7. M. Kaveh, M. Rosenbluh, I. Edrei and I. Freund, Phys. Rev. Lett. 57, 2049 (1986).
8. A. Z. Genack, Phys. Rev. Lett. 58, 2043 (1987).
9. A. A. Golubentsev, Zh. Eksp. Teor. Fiz. 86, 47 (1984). (Sov. Phys. JETP 59, 26 (1984)).
10. E. Akkermans, P. E. Wolf and R. Maynard, Phys. Rev. Lett. 56, 1471 (1986).
11. M. J. Stephen and G. Cwilich, Phys. Rev. 34, 7564 (1986).
12. F. C. MacKintosh and S. John, Phys. Rev. B37, 1884 (1988).
13. E. Akkermans and R. Maynard, J. de Physique, to be published.
14. J. W. Goodman, J. Opt. Soc. Am. 66, 1145 (1976).
15. B. Shapiro, Phys. Rev. Lett. 57, 2168 (1986).
16. M. J. Stephen and G. Cwilich, Phys. Rev. Lett. 59, 285 (1987).
17. B. L. Altschuler, Pis'ma Zh. Eksp. Teor. Fiz. 41, 530 (1985); (JETP Lett. 41, 648 (1985)).
18. P. A. Lee and A. D. Stone, Phys. Rev. Lett. 55, 1622 (1985).
19. S. Feng, C. Kane, P. A. Lee and A. D. Stone, Phys. Rev. Lett. 61, 834 (1988).
20. M. J. Stephen, Phys. Rev. B37, 1 (1988).
21. G. Maret and P. E. Wolf, Z. Phys. B65, 409 (1987).
22. D. J. Pine, D. A. Weitz, P. M. Chaikin and E. Herbolzheimer, Phys. Rev. Lett. 60, 1134 (1988).
23. I. Freund, M. Kaveh and M. Rosenbluh, Phys. Rev. Lett. 60, 1130 (1988).

Nonresonant Effects in CO_2 Amplifier of Ultrashort Laser Pulses

S. Chelkowski and A.D. Bandrauk

Departement de chimie, Université de Sherbrooke, Sherbrooke,
Québec, Canada J1K 2R1

1. Introduction

The coherent resonant interaction of laser pulses shorter than the medium relaxation times (called ultrashort pulses) has been discussed previously in detail (1967–1970) for media consisting of two–level systems and gives rise to nonlinear coherent phenomena such as self induced transparency and soliton propagation in attenuating media [1–3]. In amplifying media these phenomena allow one to obtain very short and intense pulses. Molecular systems cannot be in general considered as the two–level systems, so there is a need to examine pulse propagation in many level–systems both theoretically and experimentally.

Recently, the amplification of subpicosecond pulses in CO_2 resonant with 10.6 μm lasing transition 001↔100 has been achieved for intensities up to $I=10^{12}$ W/cm^2 [4]. Propagation of such pulses cannot be described by two–level models because of the following effects. Firstly, the subpicosecond pulse cannot be considered as a monochromatic wave; its frequency spread becomes of the order 1/(pulse duration)=33 cm^{-1}. Secondly, the power broadening effect (dynamical Stark effect) can couple many CO_2 levels at high intensities. This can be seen by considering the transition probability in the two–level system interacting nonresonantly, in the dipole approximation, with a plane wave of frequency ω. The Schroedinger equation yields, in this case, the following expression for the transition probability between the two levels [5]

$$P = \Omega^2 (\Delta^2 + \Omega^2)^{-1} \sin^2 \left[\sqrt{\Delta^2 + \Omega^2}\, t /2) \right] \tag{1}$$

where $\Omega = p\varepsilon/\hbar$ is the Rabi frequency, p is the transition dipole moment, ε is the amplitude of the electric field and $\Delta = (E_2 - E_1)/\hbar - \omega$ is the detuning. Adopting the criterion that the two levels are coupled if this probability exceeds 0.1, one concludes from (1) that a nonresonant level should be taken into account if the corresponding Rabi frequency exceeds $\Delta/3$. This allows us to estimate the importance of various CO_2 transitions usually neglected in model calculations [6,7]. These calculations are correct for low intensity long pulses. At the pulse peak intensity $I=10^{12}$ W/cm^2 the Rabi frequency corresponding to lasing transition P(20) is 7 cm^{-1} [8] which becomes close to the spacing between this transition and R(18) transition equal to 32 cm^{-1} [9]. For higher intensities, the 9.6 μm, 001↔020 transition should be taken into account, as well as, the transitions from the level 001 to levels 000, 002, 003 etc — see fig. 1. These last transitions, although characterized by the detuning as large as 1380 cm^{-1}, have the dipole moment 14 times larger than the lasing transition, leading to the Rabi frequency 980 cm^{-1} at intensity $I=10^{14}$ W/cm^2.

Springer Proceedings in Physics, Vol. 39   **Disorder and Nonlinearity**
Editor: A.R. Bishop   © Springer-Verlag Berlin, Heidelberg 1989

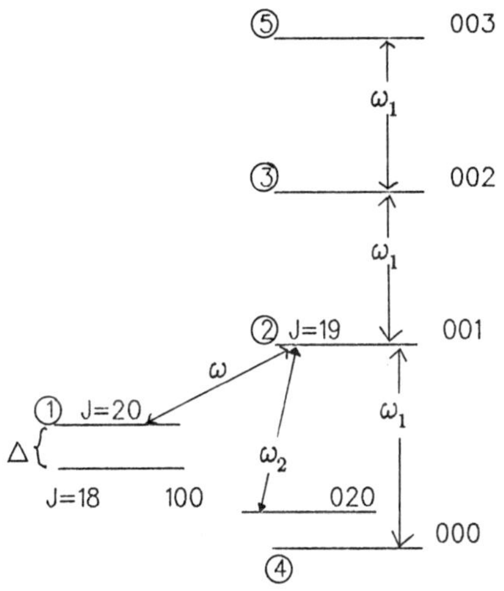

Fig. 1. Schematic diagram of the CO_2 energy levels used in model calculations. The numbers 1–5 indicate levels used in the calculations of transition amplitudes presented in fig. 5. $\omega_1/(2\pi) = 2323$ cm^{-1}, $\omega_2/(2\pi) = 1041$ cm^{-1}, $\omega/(2\pi) = 943$ cm^{-1}, $\Delta/(2\pi) = 30$ cm^{-1}

In this article we present numerical calculations which allow us to estimate the importance of these nonresonant multilevel effects.

Since the pulse duration is shorter than the medium relaxation times we expect the coherent nonlinear effects to dominate allowing us to use the Schroedinger equation to determine the transition amplitudes and the induced polarization in the medium. In the case of the two–level medium resonant with the incoming light the induced polarization is a sine function of the pulse area and the pulse evolution equation is a sine–Gordon equation [1–3]. We have shown recently that if a resonant medium consists of N–equidistant levels the induced polarization can also be found anlytically as the function of pulse area and the pulse evolution is described by <u>multisine–Gordon equation</u> [10]. In particular, in the case of such three–level medium the pulse area satisfies the double sine–Gordon equation whose solitary waves were discussed in ref. [11].

2. Maxwell–Schroedinger Equations

Supposing that N nondegenarate levels of a molecule are coupled via dipole interaction with a linearly polarized radiation field of frequency ω described by an electric field

$$E(t,z) = \varepsilon(t,z) \cos(kz - \omega t + \psi(t,z)) \tag{2}$$

where ε and ψ are the slowly varying field amplitude and phase, we get from the Schroedinger equation the following differential equations determining the time evolution of the transition amplitudes c_j :

$$i\dot{c}_j = V_{j,j-1} \exp(i\omega_{j,j-1}t) c_{j-1} + V_{j,j+1} \exp(i\omega_{j,j+1}t) c_{j+1} \tag{3}$$

where $V_{i,j} = - p_{i,j} E(t,z)/\hbar$, $\omega_{i,j} = (E_i - E_j)/\hbar$, $p_{i,j}$ are the transition dipole moments from the i–th to the j–th level. It is supposed that only adjacent levels are coupled. As a result of multiphoton transitions the induced medium polarizibility p becomes in general a complicated nonlinear functional of the pulse shape E, which is determined by the quantum definition of the polarizibility

$$p[E] = < \psi \mid \mu \mid \psi > = \Sigma \; c_j^* c_{j+1} \; \mu_{j,j+1} \; \exp(i\psi t) \; + \; c.c. \qquad (4)$$

where $\mu_{j,j+1} = < \psi_j \mid \mu \mid \psi_{j+1} >$ are the spectroscopic transition dipole moments between neighboring states. Following the formalism used for the two-level systems[1-3], we next decompose the induced dipole moment p into parts in phase and out of phase with the propagating electric field $E(t,z)$, i.e.

$$p = p_{12} \; (\delta \; \sin\Phi + C \; \cos\Phi) \quad \text{where} \qquad \Phi = kz - \omega t + \psi(t,z)$$

is the total spatial temporal phase. Next one assumes a slowly varying envelope approximation for ε, ψ, δ and C with respect to the free wave part $\exp[i(kz-\omega t)]$. Thus the Maxwell's wave equation in polarized medium is

$$c^2 \; (\partial^2 E/\partial z^2) - (\partial^2 E/\partial t^2) = 4 \; \pi \; n_o \; (\partial^2 p/\partial t^2) \qquad (5)$$

where n_o is the number of particles per cm^3 ,which leads to the following form of the matter field equations

$$(\partial\varepsilon_c/\partial t) + c \; (\partial\varepsilon/\partial z) = 2 \; \pi \; n_0 \; \omega \; p_{12} \; (\delta + iC) \qquad (6)$$

where $\varepsilon_c = \varepsilon \; \exp(i\psi)$ is the complex slowly varying field envelope.

3. Pulse propagation in three-level systems

We have integrated numerically the equations (3) and (6) for the case when an optically active medium contains three level systems presented in Fig.2 in order to mimic by a simple model complex multilevel transitions in CO_2 . The nonresonant level 3 induces virtual transitions.

The input pulse at z=0 was chosen to be a pulse having the total area

$$S = (p_{12}/\hbar) \int_{-\infty}^{\infty} \varepsilon(t,z) \; dt \qquad (7)$$

equal to $\pi/2$. The resulting pulse shapes $\varepsilon \cdot p_{12}$ in cm^{-1} at $z = 598$ and $z = 1095$ cm are shown for detuning corresponding to R(20) transition in Fig.3a and to $001 \leftrightarrow 020$ transition in fig. 3b . Both should be compared with Fig.4 describing pulse propagation in a simple two-level medium.

One concludes that at the intensity $I = 2.35 \cdot 10^{13}$ the contribution of the $001 \leftrightarrow 020$ transition to the pulse amplification is already noticeable (fig. 3a) and that the R(20) branch enhances substantially the pulse amplification fig. 3b)

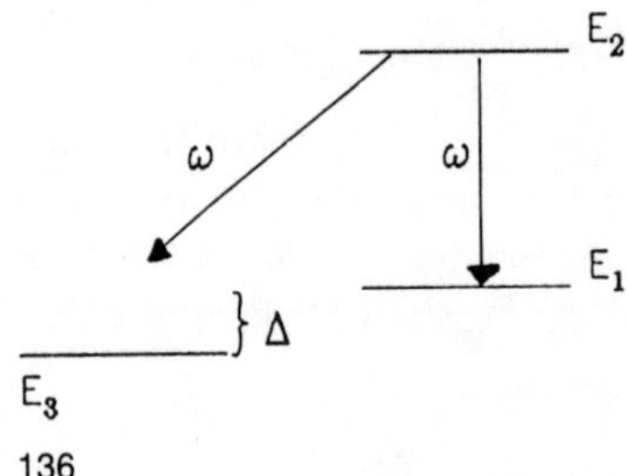

Fig. 2. Schematic diagrams of energy levels used in model calculations presented in fig. 3. The level E_1 and E_2 represent the lasing levels 001 and 100 for J=19 and 20 correspondingly.

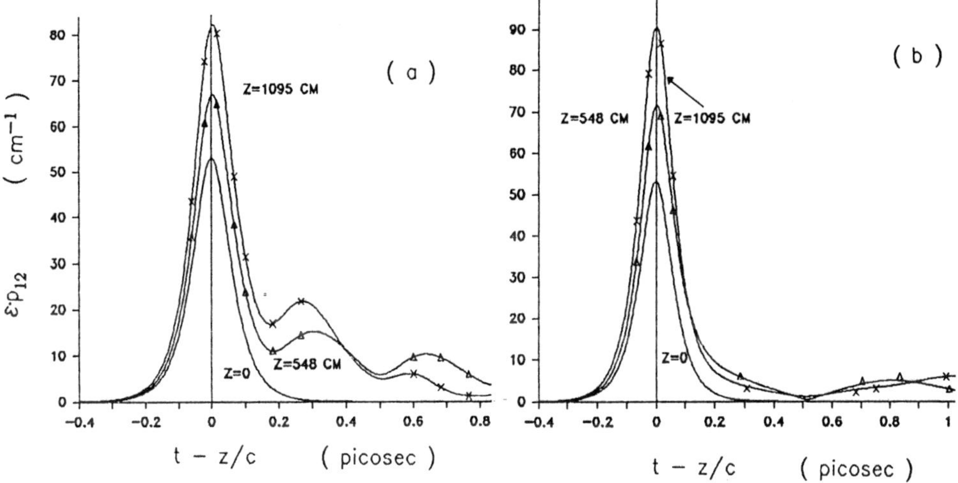

Fig. 3. Pulse shapes $\varepsilon\cdot p_{12}$ [cm^{-1}] for a level scheme of fig.2. Transition dipole moments $p_{12} = p_{23} = 0.015$ Debye, $n_o = 5.85 \cdot 10^{17}$ molecules/cm^3, input peak intensity I = 2.35 $\cdot$ 10^{18} W/cm^2 , detuning $\Delta/(2\pi)$ is a) 98 and b) 32 cm^{-1}

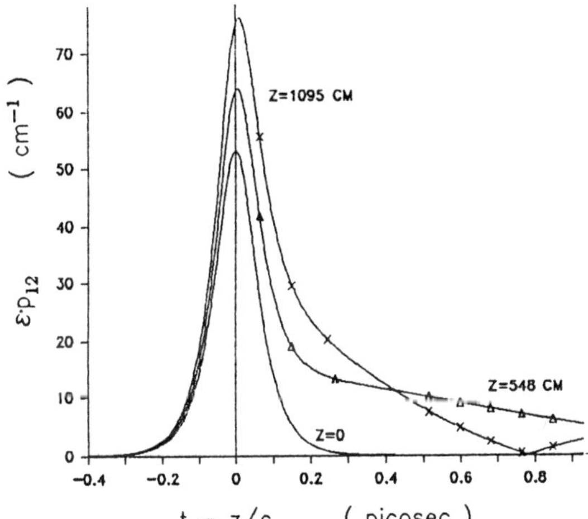

Fig. 4. Pulse shapes $\varepsilon \cdot p_{12}$ in cm^{-1} for the two-level model with the same p_{12}, n_o and I as in fig. 3.

4. Transition probabilities to higher vibrational levels

In order to evaluate the intensity at which the transitions to distant vibrational levels such as 001, 002, 003 etc should be taken into account we integrated numerically the Schroedinger equation (3) in which we included five CO_2 levels indicated bu numbers 1–5 in fig. 1. The resulting transition probabilities from level 2 to levels 1,3,5 are shown in fig. 5.

Since the input pulse area was $\pi/2$ the value of $|c_1(+\infty)|^2$ in the two-level model including only lasing transitions should be $\sin^2(\pi/4) = 0.5$. This number should be compared with the values of $|c_1(+\infty)|^2$ equal to 0.33 and 0.13 (see figs. 5a and 5b).

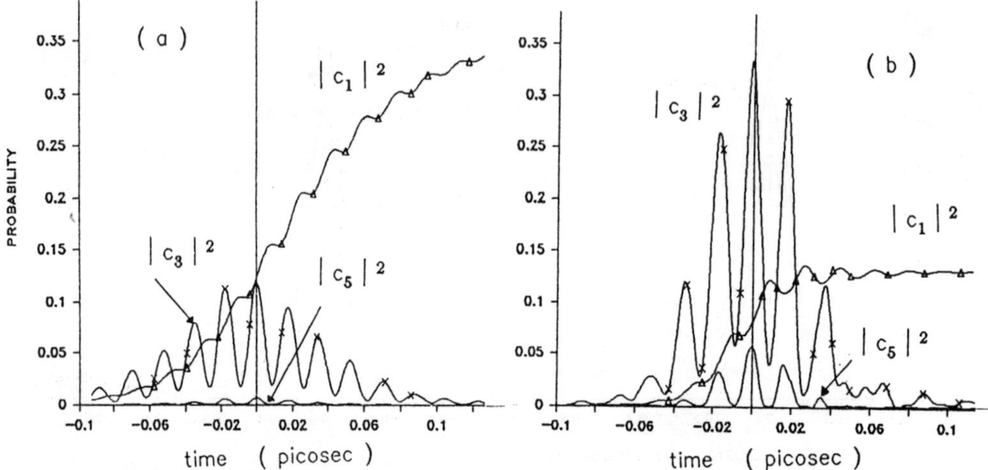

Fig. 5. Transition probabilities $|c_3|^2$, $|c_5|^2$ and $|c_2|^2$ as functions of time for $\pi/2$ pulses at intensity a) $I = 2.35 \cdot 10^{13}$ W/cm^2 ; b) $I = 9.3 \cdot 10^{13}$ W/cm^2. Initially the system was in level 2.

This means that the transitions to higher vibrational levels decrease considerably pulse amplification at intensities above $I = 10^{13}$ W/cm^2 .

We conclude also that at such high intensities the slowly varying envelope approximation cannot be made since, as seen in fig. 5, the transition amplitudes and consequently the polarization envelopes oscillate rapidly, which means that the pulse evolution should be determined by direct integration of the second order wave equation (5) together with the equations (2) without invoking the Rotating Wave Aproximation (R.W.A.). We intend to perform such integration in the near future. The R.W.A. was used in the calculations presented in the previous section but was not invoked in the calculations described here.

5. Concluding remarks

Our numerical calculations based on simple multilelevel models show that the usual two—level models, as well as, other approximations usually used in the description of resonant pulse propagation break down above the intensity 10^{13} W/cm^2 when much more complex models should be used.

However we note that the curves presented in fig. 3 cannot be directly compared with the experimental data since the level structure of CO_2 is much more complex than the structure from fig. 2 . Firstly, one should include all the transitions P(J) and R(J) with J distributed according to the Boltzman distribution. Only at very low intensities does the transition P(20) dominate; the contribution of other P(J) transitions gradually increases with increasing intensity. Secondly, if the pulse area is comparable to $\pi/2$ the level degeneracy must be taken into account, since because of nonlinear effects it is no longer correct to include it by taking just the averaged value of the transition dipole moment [2,6]. We will present in the near future the results of the calculations which include all these effects [12].

Acknowledgements

We thank the Natural Science and Engineering Research Council of Canada for financial support. We also wish to acknowledge informative discussions with Dr. P. Corkum on high intensity pulse amplification in CO_2 .

References

1. J. Allen, J.H. Eberly, <u>Optical Resonance and Two−Level Atoms</u> (J.Wiley Sons Inc., N.Y. 1975)
2. G.L. Lamb, Jr., Rev. Mod. Phys. <u>43</u>, 99, (1971)
3. G.L. Lamb, Jr. <u>Elements of Soliton Theory</u> (J.Wiley Sons Inc., N.Y. 1980)
4. P.B. Corkum, IEE J. Quant. Electron. QE−21, 216 (1985); in <u>Laser Acceleration of Particles</u>, AIP Conf.Proc., #130, 493 (1985)
5. J.I. Steinfeld, <u>Molecules and Radiation</u>, (Harper Row Publishers, N.Y. 1974) p.271
6. A.O. Markano, V.T. Platonenko, Sov. J. Quantum Electron. <u>10</u>, 433, (1980)
7. V.T. Platonenko, V.D. Taranukhin, Sov. J. Quantum Electron. <u>13</u>, 1459, (1983)
8. We defined in this paper the Rabi frequency as a product of the electric field intensity and the transition dipole moment, averaged over magnetic quantum number m, corresponding to P(20) transition equal to 0.015 Debye [13−15]
9. G.H. Herzberg, <u>Molecular spectra and Molecular Structure</u>, Vol.2, Van Nostrand Reinhold Company, N.Y. (1945), p.14 and 21
10. S. Chelkowski and A.D. Bandrauk: In <u>Atomic and Molecular Processes with Intense Laser Pulses</u>, ed. by A.D. Bandrauk, NATO ASI Series, Series B:Physics, Vol.171 (Plenum Press, N.Y. 1988), p.57; J. Chem. Phys. <u>89</u>, 3618 (1988)
11. D.K. Campbell, M. Peyrard, P.Sodano, Physica <u>19D</u>, 165 (1986)
12. S. Chelkowski, A.D. Bandrauk, submitted to J. Opt. Soc. Am. B,
13. A.M. Robinson, Can. J. Phys. <u>50</u>, 2471 (1972)
14. I.B. Burak, L.A. Gamss, J. Chem. Phys.,<u>65</u>, 5385 (1977)
15. A. Yariv, <u>Quantum Electronics</u>, (J. Wiley Sons Inc.,N.Y. 1975) p.550

Part IV

Postscript

Disorder and Nonlinearity

J.A. Krumhansl

Cornell University

These brief remarks paraphrase my extemporaneous comments, when asked to summarize in closing the workshop.

Three significant areas of research (at least) have been represented: first, several mathematical analyses of soluble nonlinear models in which the introduction of disorder modifies familiar behavior in essential ways; second, discussions of "weak" localization, now being looked at not only in the random lattice electron problem but from a general wave scattering viewpoint, indeed experimental evidence for optical localization was presented; third, problems in strong but not complete localization where the asymptotics of transport are not yet understood. Various computer simulations supported the presentations.

My comments are to the effect that to a good approximation one might describe the meeting as two interesting workshops, one "Disorder" and the other "Nonlinearity." Indeed, a *a priori*, there need not have been a connection. However, I think that the organizers did have in mind the possibility of finding some connections. If that did in fact come to pass, a different title would be then more accurate. "Disorder <u>with</u> Nonlinearity." Is there a difference? One of my eminent teachers clearly thought so, and pointed out that a "woman and child" are rather different from a "woman with child." In fact there is hope that the two cultures which met at Los Alamos are beginning to find common ground. Maybe, then, this workshop will have given birth to a new mutation embodying both nonlinearity with disorder in an intrinsic way. No doubt a subject for future workshops.

Springer Proceedings in Physics, Vol. 39   **Disorder and Nonlinearity**
Editor: A.R. Bishop   © Springer-Verlag Berlin, Heidelberg 1989

Index of Contributors